THE BIG BANG
and
LINES OF SPACE
A Simple Dialogue on Two
Diverse Concepts

Authored by
Devinder Kumar Dhiman

Registered with the IP Rights Office
Copyright Registration Service
Ref: 3719293366

License

This book is licensed for your personal reading only. This book may not be re-sold or given away to other people. If you would like to share this book with another person, please purchase an additional copy for each recipient. If you're reading this book and did not purchase it, or it was not purchased for your use only, then please return to www.amazon.com and purchase your own copy. Thank you for respecting the hard work of this author

Dedicated To My Colleagues Who Gave Me Valuable Support

Introduction

13.7 billion years ago, an infinitesimal point in space exploded with such a gigantic force that it is still expanding. During its expansion, it caused the formation of billions of galaxies, numerous stars and our own solar system, including the Earth. This idea of creation of the universe is known as the Big Bang, and this is the most prevailing scientific concept of the creation of the universe. But this concept has some weaknesses. Therefore, a perpetual hunt for a better concept is on. There are many alternative ideas being put forward. One of them is "Lines of Space".

"Lines of Space" is imagined to be a medium of space similar to the Ether that was proposed in the 1860s by the famous scientist James Clerk Maxwell, who had also proven that light is an electromagnetic radiation.

The book elucidating the concept of Lines of Space is written in a conversational style and has got very good remarks from the readers. A few of them are written after this Introduction, just to give an idea what the readers have said about *Lines of Space*.

After writing that book, the Author encountered a few incidents where he became aware that, in general, there is not enough clarity in the understanding of the Big Bang theory. So, he decided to write this book and give a comparison between the Big Bang, and Lines of Space. This book also has been written in the same conversational style as the first book *Lines of Space*, but with a different background. Although it is recommended to read the first book also, it is not a must for reading the present book, which is self-explanatory in all aspects.

A Few Reviews of *Lines of Space*

"Absolutely fascinating concept! I would like to see the scientific community explore this more. The book presents the ideas in an easy to follow manner and is as entertaining as it is thought provoking. I have recently been recapturing my love of physics and reading about quantum theory and other interesting matters, so when I came across this book I had to explore it. I highly recommend this to anyone who has a curious mind and enjoys pondering the Universe!"
~ **Chris Bourlier (Acienda Heights, CA, US)**

"This book is written in the context of family, which makes it so much more interesting. It is an exploration of what could have caused the beginning of everything and is well thought out. What it needs now is the attention of physicists to do their own calculations to test the author's theory and see if it is relevant and leads to more understanding. After all, that's what science is supposed to do. It makes sense to me, but I'm only an armchair scientist - I enjoyed reading about it, but having CLEPed the sciences in college I hope this gets the attention of scientists so I can read about their deductions and assumptions."
~ **Allison Kohn (Colorado) – Author**

"The author has done a very good job of breaking down the concepts of physics and making it easier for young people and the ordinary layman to understand how the Universe works. Starting from the inner workings of the atom itself, the author leads us through a simplified discussion of physics that eventually leads us to his own unique theory of an unseen medium in space in the form of "lines of space". From what I understand, these lines are similar to the strings in string theory that we hear so much about today from scientists but with specific properties. From here the author delves into formulas and examples of how these lines of space can be

detected. Mr. Dhiman presents a very compelling argument for his theory and it will be interesting to see how this plays out eventually in the scientific community. The illustrations presented throughout the book help the reader to understand the topics discussed."

~ Lee MacRae – Author

A Comment on *The Big Bang and Lines of Space*

After reading the author's first book 'Lines of Space', I was very much interested in 'The Big Bang and Lines of Space' as my curiosity was raised. In this book, he has revealed the numerous inconsistencies in The Big Bang theory through a very interesting conversation. This left me wondering at what I knew the Big Bang was, 'an explosion of a small point', which we are used to and believe in. After reading it, I find that the alternate idea of the creation of the universe put forward by the author is very rational. Could this really be true. Maybe. One needs to read the book to argue in favour or against. All I can say is keep an open mind before you make any conclusions. Interesting reading even for people without any scientific background.

~ **Sunil Shetye (Master Mariner, Pune, India)**

A Word from the Author

This book is not restricted to the conventional belief of physics that space has no medium. It explores the possibility that empty space could also contain some kind of medium. This medium is considered as 'Lines of Space,' and it is conjectured that the matter could have been created from this medium, instead of creation by the Big Bang. During the course of proving this assertion, I had to deviate from the prevalent theories of physics. Therefore, the reader may find in this book such notions that are different from the recognized theories of physics. My intention is not to challenge any reputed theory of science. I only wish to put forward my viewpoint.

*****
"If the idea at first is not absurd, then there is no hope for it"
~ Albert Einstein

******

Table of Contents

Part I – The Big-Bang

Chapter 1 – At the Airport	11
Chapter 2 – Nandi Bull	17
Chapter 3 – Meeting at the Airport	21
Chapter 4 – On the Flight	23
Chapter 5 – Red Shift	27
Chapter 6 – Expansion of the Universe	35
Chapter 7 – C.R.M.B	41
Chapter 8 – Abundance of Elements	47
Chapter 9 – Inflation Theory	51
Chapter 10 – God Particle	53
Chapter 11 – At the Guest House	57
Chapter 12 – Ptolemy's Epicycles	61

Part II – Ether

Chapter 13 – Ether	65
Chapter 14 – Perihelion of Mercury	79
Chapter 15 – Bending of Light	89
Chapter 16 – Refraction of Light	101

Part III – Lines of Space

Chapter 17 – Lines of Space	107
Chapter 18 – Sub-atomic particles	119
Chapter 19 – A Comparison	123
Chapter 20 – The Leave Taking	127
Acknowledgment	129
Declaration	132

PART I
THE BIG BANG

Chapter 1
At the Airport

After passing through the security check at the Chandigarh Airport, I proceeded to the waiting lounge. I paused at the Information screen and learned that my flight would be delayed an hour. I frowned. Although it had taken me quite some time to check in and get through the security check because of the unusually large holiday crowd, I had still managed to reach the waiting lounge an hour earlier than my scheduled departure time. Now my wait would be considerably longer.

I surveyed the crowded lounge for an empty chair. Having finally spotted one, I excused my way past an elderly lady and a middle aged man to get to it. As soon as I sat down, the man beside me struck up a conversation by commenting on the unusual rush on a Sunday morning.

I agreed, and the conversation progressed. He told me that he was an engineer who worked in Mohali, a nearby town, and was going back home to Mumbai for a two week holiday.

"Do you work in Mumbai?" he asked as he guessed that I was a local and must be going to Mumbai for work.

"No. Just going to attend training for one week," I told him.

"Working in IT or insurance?" he asked as he tried to make a second guess.

"No. Shipping," I corrected him.

"Okay," he acknowledged, then grew silent. I took the opportunity to ask him whether he would be interested in reading my book. I had recently published my first book and was declaring the fact to almost everyone who came my way. As he was an engineer, and my book dealt with science, I reckoned him to be a good candidate for giving me feedback.

He asked, "What is the subject?"

"It is about a new concept for the creation of the universe."

"I know about the creation of the universe…the Big Bang theory."

I became hopeful that the man would be interested in my book, but contrary to my hope, he said, "There is nothing more to know about that."

Not ready to give in just yet, I quickly added, "My book is about a different concept on the creation of the universe."

He replied with certainty, "I've known of the Big Bang theory since my early days; there can't be any other reason for the creation of the universe."

He was obviously a strong believer of the Big Bang theory, and I needed to tread very carefully if I wanted to avoid infuriating him and losing any chance of getting him interested in reading my book. To continue the conversation with him, I asked him politely if he would be kind enough to explain the Big Bang theory, as he understood it.

He replied, "It is very easy; millions of years back, two planets collided and they caused a big bang, thus creating more planets and stars. This way, the entire universe was created."

I was dumbstruck at his answer and did not know whether to attempt correcting him or just stay indifferent. From his point of view, he was right as he gave the explanation, for which he did not require any further persuasion.

He continued, expanding on his statement,"You know…it's how everything gets created; two living beings meet and they reproduce a new being. Similarly, there must be two planets meeting and forming a younger third planet and hence their population also could keep growing. This is the Big Bang theory."

I opted to stay quiet for awhile and wondered at the difficult task that lay ahead of me if I were to convince him about the incorrectness of his long-held belief in his own kind of Big

Bang theory.

"How did those first two planets come into being?" I finally asked.

"They were there from the beginning of the universe itself," was his instantaneous reply, but my question left him wondering about it, as could be observed from his face. I felt certain nobody had ever posed such a question to him. I waited for him to speak instead of resuming the conversation, giving him plenty of time to reconsider the notion he'd probably learned as a child.

"Yes, you are right," he finally said. "Where did those two planets come from in the beginning? They couldn't have been there from the beginning of the universe."

"Right. If there were not even two planets in the beginning of the universe, they couldn't have collided, and then there couldn't have been a Big Bang," I carefully added and looked at him for a response.

"Then what was the Big Bang?" he asked me.

I was surprised at the ease with which I could make him understand that his notion of the Big Bang was incorrect, but it was his capacity to think as a logical person that helped to dispel his wrong notion more quickly than anything to do with my own knack of teaching.

I clarified the current popular theory to him, "The Big Bang theory is not really about a big, loud, bang of two massive objects as you probably assumed it to be, from the words 'Big Bang'. The name is a misnomer: the theory does not imply a collision of two big objects. It is an explosion of a very small point-like particle that caused the creation of the universe."

"That is more improbable. How can one small point explode to form the universe?" he disagreed with me.

"You are right. It is not possible that the universe could have been created either from the colliding of planets, as you believed formerly, or by the explosion of one small point-like object into billions of stars and galaxies, as the Big Bang theory advocates," I agreed with him.

"Then how did the universe get created?" he asked.

Before I could respond, there was an announcement that the flight had been cancelled. There was suddenly a chaotic scene. People were asking one another, "What happened?" Nobody had a clue for a few moments until a second announcement was made that the plane had developed a snag and it would have to go for repairs. The announcer expressed deep regrets and assured all the passengers that an alternative arrangement would be made soon.

"What is this?" the man sitting beside me asked disgustedly, whereas the elderly lady sitting on the other side seemed worried. She grumbled, "My son has already left after seeing me off. What do I do now?"

"Don't worry, Madam; the airline will make some arrangement. We should wait." I assured her, while I noticed some passengers becoming restless and blaming the airline for canceling the flight at the last minute.

Seeing the commotion, the airline staff came to pacify them. An official of the airline tried to explain the situation, but things were getting out of control. A security guard came to her rescue and asked the people to listen to her. She explained that the airline would provide a road transport to the Delhi airport, and from there they could board the next flight to Mumbai.

"Why not a flight from here?" asked someone.

"We operate only one flight from here daily," she replied and then added, "But we can connect from Delhi."

"Why not make the adjustment with other airlines?"

"They are all full."

"If we don't want to go by road transport?" asked another passenger.

"Then you can either take a refund, or you can travel tomorrow," she clarified.

"Where will we sleep tonight, here?" asked a South Indian man, who probably didn't have a place to stay in Chandigarh.

"We'll provide you a hotel stay." The airline official seemed to be in full control of the situation and had the authority to make all decisions necessary to handle the situation.

"When will the transport leave for Delhi?" someone asked her.

"In two hours from now. It will reach Delhi by 6:00 p.m. We have a flight for Mumbai departing at 9:00 p.m. from Delhi," she said.

I made a mental calculation that if I took the flight the next morning, then I'd reach Mumbai in the afternoon and would miss the first day of my training. This was not a good idea considering my company was going to pay a hefty sum of 4000 USD for my one week of training, in addition to the expenses incurred on my travel, food, and lodging. No, I couldn't postpone my travel by one day. Lost money for the company, and a day of learning lost for me was unacceptable. But a flight at 9:00 p.m. from Delhi also would not be convenient as I would reach Mumbai at midnight. I was in a fix, but I had to make a quick decision.

It was Sunday, so I hesitated disturbing my company manager in Mumbai, but I had to inform him about this sudden change. I called him up. After listening to me, he told me to wait for his call while he tried to make some arrangement online.

In the meantime, other passengers had started informing the airline staff about their preferences. After five minutes, I got the instructions from my company manager to hire a taxi and reach Delhi airport within six hours as he was going to book the flight departing at 4:15 p.m. from Delhi.

I hurried through the security gate after informing the airline staff that I was going on my own, and the refund would be handled by my company. I collected my baggage and rushed towards the taxi counter.

The Big Bang and Lines of Space

Chapter 2
Nandi Bull

The taxi was on the highway to Delhi within five minutes of departure from the airport thanks to light Sunday morning traffic. I was seated comfortably in the back seat of the taxi when my mobile rang.

"Hello," I said into the mobile.

"Hi, where are you…In the plane?" came the voice of my wife on the other end.

"No. In a taxi."

"What? I had dropped you at the airport, not a taxi stand."

"Yeah."

"Then…Oh! You mean the plane is taxiing?"

"No. I am in a taxi going to Delhi," I replied.

"Why? What happened?" She sounded concerned. I told her the whole story and disconnected the phone after saying goodbye.

Then I realized that I didn't even say goodbye to the man with whom I was talking to at the airport. I had left in a hurry. It was not that it mattered much now, but it would have been nice to say goodbye; after-all we had talked for quite some time. I had become speechless momentarily when he had told me about his idea of the Big Bang, but on taking a second look, I realized it was probably not his fault.

I remembered the day, nearly four decades back, when my grandfather was having his lunch and I was standing by, ready to serve him more vegetables or *chapattis*. He sat on a rug spread out on the ground. There was only one table and two chairs in our house, but they were specifically used only when guests arrived. Otherwise, everyone sat on the ground while eating their food.

I must have been only nine or ten years old at that time. I

was keeping an eye on the plate in front of him so I could bring more *chapattis* as soon as he needed them. Suddenly, the plate started moving away from him as if a ghost was pulling at it. He tried to pull the plate back towards him, but it moved away again.

He abruptly stood up and shrieked, "Earthquake!, Run! Come out!"

I was spellbound and did not know what was happening. He caught hold of my arm and pulled me along with him, dashing out of the room while holding the folds of his *dhoti* (lower garment) with his left hand, and continuously shouting for others to come out of the house. Within a few seconds, I saw my family members and the neighbors rushing out of their homes into the street. A crowd had started gathering in the open area in front of our house.

Our house was in the center of the street, so the open space in front of our house was the usual meeting point for all our neighbors. Young ones were bewildered at what was happening whereas the elders did seem to understand and were guiding the others. There was a sense of urgency as they accounted for their loved ones, making sure no one had been left in the houses.

That had been my first encounter with an earthquake. We were lucky that none of the houses had fallen and nobody had been hurt. As the realization dawned on the people that the worst had passed and they were safe, they relaxed. Some of them started telling their stories, what they were doing when it started. Someone said, "My cot started shaking, and I asked my brother why are you doing that? Then I heard the shouts of earthquake." A woman said, "My overhead fan started swinging."

Within minutes the scary situation had turned into a cheerful one, as the people realized that they had not incurred any losses. They were blissfully narrating their incident as if the earthquake was a play thing.

An old lady said, "It has come after nearly thirty years."

"Why did it come?" asked a young man who had gotten

married recently.

"Nandi changed his horn," the lady replied.

"Nandi?" he asked and looked at the man standing a few feet away. Nandi was the grocery store owner. How could he do something like this? It was beyond his comprehension. Not only he, but many others looked bewildered.

"No. She does not mean this Nandi; she means God Shiva's Nandi. His bullock," my grandpa clarified.

"Oh! Nandi Bull." Some of the elder people understood and lowered their head in reverence.

"Yes, Nandi is the bullock, the vehicle of God Shiva, and he holds the Earth balanced on his one horn. When he gets tired, he changes it to the other horn, that causes an earthquake," explained the old lady, and everyone listened intently. It was not only believed by her, it was quite a common anecdote there, which was apparent as the other elders agreed with her immediately.

I kept wondering for several days why Nandi could not carry the Earth on both of his horns so that he wouldn't have to shift it from one to the other, and earthquakes could be avoided. Even better would be to carry the Earth on his back as normal bulls do when they carry a load. But I was not to question the ways of the Lord, as was instilled into me. I was simply told to believe it. It was only after two or three years, after I had read about seismology in a school text book, that I understood the real reason earthquakes occurred.

That incident taught me how mislaid the beliefs of the people could be. Thus, the mistaken belief of the man at the airport was no wonder. It is only education that can set things right. But our education also has such misconstructions. For example, in the case of the Big Bang, how can a point-like particle suddenly burst into stars and galaxies? It seems like a fallacy. Not only religious beliefs, but also scientific studies, sometimes appear to have no reason. I would have kept pondering it over, but a drop in the speed of the taxi pulled me out of my deliberations. I propped myself up from the reclining posture to find out the reason. A toll barrier lay ahead of us.

After paying the toll, I informed the driver that we had to reach the airport before 3:00 p.m.

"What time is your flight, sir?" he asked.

"4:15 p.m."

"The highway is no problem, but I can't say about passing through Delhi. That can delay us." He expressed his apprehensions.

"You will get two hundred rupees extra if we reach the airport before 3:00 p.m." I offered. I suspected that he was not the owner of the taxi and could only be getting about four hundred rupees (Approximately seven USD) per day. An addition of fifty percent to his daily wages would be attractive for him.

"You will reach there before that," he said with a grin. The additional offer had the desired effect.

"But drive safely," I warned him.

"Don't worry, I will not over-speed. We'll cut down the break time of about half an hour. I will take a rest only after I drop you at the airport," he said making me aware of his plans.

"That will be good," I approved.

Chapter 3
Meeting at the Airport

The taxi stopped at the domestic airport terminal of Delhi after a non-stop run of five hours since the driver had not taken his customary break mid-way. I paid the driver along with the extra money that I had promised to him. He gladly accepted it, quickly took out my luggage, loaded it onto a trolley, and bade me a happy journey. I almost ran towards the entrance of the airport, but slowed and felt a wave of relief when I noted the check-in counter open, and only a few passengers waiting to check-in.

"Good afternoon, sir," I heard a somewhat familiar voice say while I was heading towards the check-in counter. I turned in the direction of the sound and found Gaurav.

"Hey, how are you?" I was surprised to see him there. He had worked as an assistant engineer under my supervision about seven years back. Then, he was a young boy fresh from the engineering college. Now, he was in his early thirties, looking more confident and handsome, but his voice had not changed much. I was quick to recognize him after so many years.

"Fine, sir. I have been waiting for you. You were delayed."

"Yes. But how did you know that I was coming? What are you doing here?" I had lost touch with him six years ago after we both left the ship we had been working on together. We now sailed on separate ships.

"First, let me help you with check-in, then we shall talk," he replied and guided me towards another counter that had become vacant.

He picked up my suitcase from the trolley and put it up for check in, while I told the female airline attendant to check for my online booking, as I had no time to print my e-ticket before reaching the airport. She asked for my identification, and checked her computer. She found my e-ticket and asked for any seat preference.

"Next to mine, if possible," Gaurav replied before I could say anything, and gave his seat number to her.

"Done. Have a nice journey, sir," the attendant said as she handed me my boarding pass.

"Sir, I was told by the Mumbai office that your flight from Chandigarh to Mumbai was canceled and you shall be traveling with me from here to Mumbai, so I was waiting for you." Gaurav informed me. "Let's go for the security check."

"Why are you going to Mumbai?" I asked.

"I shall be attending the same training course as you," he replied and headed towards the security gate.

"What are you sailing as?" I asked him after clearing the security check.

"I am presently second engineer, but the company has promised me a promotion soon, and I expect to go on the next ship as Chief Engineer."

"Congrats!" I appreciated his fast rise from assistant engineer to the highest rank as engineer on a ship in such a short span.

"Thank you, sir. I have been sailing longer than I stayed at home, so I had good sailing experience in a short time, and got faster promotions than my colleagues," he explained and kept talking until we reached the boarding gate area.

"We have about ten minutes before boarding. I shall have a coffee. Are you coming with me?" I asked after we had found two empty seats in the boarding area.

"No, sir. I am good. You go ahead."

"Then please watch my bag." I left my carry-on on a seat next to him and ambled towards a coffee shop, fully trusting a man whom I had met again after nearly six years.

Chapter 4
On the Flight

Gaurav had gotten the window seat and I sat next to him. It was nice that he had waited for me at the check-in counter so we could get adjacent seats. Talking to him would help pass the flight time, I thought. It becomes boring during short flights due to non-availability of a video screen in front of your seat, unless you have a book to read or the company of someone. The aisle seat next to mine was vacant, although the rest of the seats were almost full.

I asked Gaurav about what he had been doing the past six years. He told me the names of the ships on which he had worked, and when he had cleared his examinations for promotion. He had a son who was now one year and seven months old. He kept on talking until the plane reached the runway and got ready to take off.

As the plane took off, Gaurav started peeking through the window and looked like he was observing the buildings of the city underneath, which appeared smaller as the plane kept gaining height. He became quiet as the panorama outside seemed to entrance him. I wanted to tell him about my book, but restrained myself from disturbing him. I had a hunch that he would resume the conversation as soon as the scenery outside faded out of his sight. But he kept observing the blue sky, seemingly fascinated by the white clouds above which the plane was flying.

There was nothing else to hold my attention there, so I pulled out the flight magazine tucked into the pocket of the seat in front of me. After a few minutes, while I was poring over the pages of the magazine, Gaurav decided to break away from his sky watching spell and spoke up. "Sir, whenever I fly, I become conscious of the vastness of the sky. In comparison to sky, the people living on Earth are tiny…even multi-storied buildings look like toys from here…It is always mesmerizing to watch."

"Yes, it is," I said without taking my eyes off the magazine.

"Sometimes I wonder how spacious the sky must be; it is intriguing to just think of the vastness!" he conveyed his thoughts to me.

"Yes, it is," I agreed.

"I've heard there are billions of stars in the universe," he expressed further.

"Yes, there are," I concurred with him and closed the magazine to give more attention to the conversation.

"Would you like to buy some snacks?" the air hostess asked as she approached our seats.

"Yes, one veggie sandwich and tea please," Gaurav requested, whereas I opted for one juice.

I opened the tray on the backside of the seat in front of me to set my juice on and made the payment for both of us, as per the Indian custom that the senior person should pay.

After the air hostess left, I asked Gaurav, "Do you know how it was created?"

"What?"

"The Universe."

"By the Big Bang." He gave the more or less standard reply.

"What do you know about the Big Bang?" I found myself asking someone about the Big Bang for the second time in one day. I had gotten a vague answer earlier, so I wanted to know what Gaurav had to say.

"The Big Bang means that the universe was created from a tiny particle due to a huge explosion," he replied correctly.

"You are correct. But it does not appear rational to me," I stated.

"Why? This is the scientific theory of the creation of the universe. Everyone believes that."

"I doubt that."

"But it is so popular," Gaurav countered.

"Popularity of something is not a yardstick of its truthfulness," I said philosophically.

"That is right, but you must have some reasoning to have doubts on such a popular theory."

"Yes, I have. It is a long story."

"Tell me. You are saying something contradictory to what I have read and believed. It will be interesting to hear."

"To explain this, I shall have to start from the beginning." I hesitated as I was not sure whether he was truly interested or merely asked in order to pass the time.

"We have enough time. I want to hear it," he expressed his interest clearly. I felt like telling him about my book *Lines of Space,* but stopped myself. I thought it might be better to first talk about the concept of Lines of Space and create an interest, rather than telling him straightaway to read the book, and decided to answer his questions.

The Big Bang and Lines of Space

Chapter 5
Red-Shift

"First of all, we shall have to understand why the Big Bang is believed to have happened," I suggested.

"Because of the expansion of the universe." Gaurav gave the prevailing reason.

"Right. But why is it claimed that the universe is expanding?" I put up a question to him.

"That I don't know," he answered honestly instead of making any guess.

"This is believed to be due to the red-shift of the light coming from stars," I told him.

"What is red-shift?"

"Red-shift is the shifting of the spectrum of light towards red color. As in a rainbow, the light splits into different colors; similarly, a light beam can be made to split in the laboratory, and then can be studied after taking a print. The print is known as its spectrum. You must have studied about this in school," I tried to remind him. He nodded weakly, which indicated that he had probably forgotten about spectrums, so I took out my laptop and showed him a photo indicating the red-shift and blue-shift of the spectrum of light.

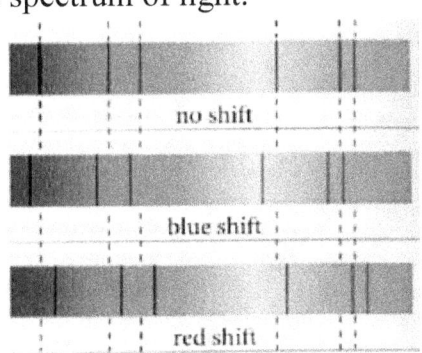

Figure 1 – Hydrogen Spectrum

"What are these vertical lines?" he asked.

"When light is passed through a gas such as hydrogen, then the light photons (light particles) of certain frequencies (*frequency is the number of wave cycles completed in one second*) get absorbed in that gas. These lines show the frequencies at which the light has been absorbed by the gas, and therefore, it is called the absorption spectrum, and will look like the top-most spectrum with no shift. But the light coming from stars, which also passes through the hydrogen gas, as is evident from the similarity in spacing of the lines, is found either blue-shifted or red-shifted," I explained.

"Why are there so many lines?" he asked.

"Do you remember the structure of the hydrogen atom?" I asked him another question instead of replying to his question.

"Yes, it has got one proton in the nucleus and one electron revolves around that."

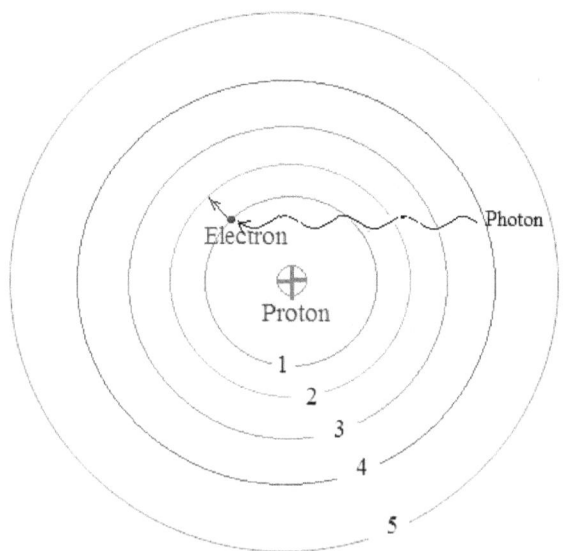

Figure 2
Electron jump between orbits

I agreed and made a simple diagram on a piece of paper, as shown in Figure 2 and explained, "When a light photon hits the electron of a hydrogen atom, it pushes the electron into higher orbit. Since the electron cannot stay in the intervening space between consecutive orbits, it absorbs only those photons of light which are able to push the electron to an exact higher level. Pushing from 1st to 2nd orbit gives one line print on the spectrum, pushing from 1st to 3rd orbit gives another line print and so on. Thus there are many absorption lines in the spectrum," I replied, effectively answering his question related to the number of lines.

"And why will the lines shift to the blue side or red side?" he asked after he understood the formation of lines in the spectrum.

"Looking at the diagram, you can see that if the distance between consecutive orbits is altered for any reason, the electron will absorb a different frequency photon, and should leave a print at a different location in the spectrum."

"Yes."

"If the energy difference between consecutive orbits is increased, then the electron will absorb a higher frequency photon (*Energy of a photon is directly proportional to its frequency*), causing the spectrum line to shift towards the blue color, and if the energy difference is reduced, then towards the red color," I explained.

"Why will this energy difference between orbits alter?" he asked further.

"It can change if the atom as a whole is either compressed or expanded. The atom may get compressed under very high gravity and may expand in extremely high temperatures. Thus, these atoms of hydrogen may behave differently in extremely high gravitational fields as in the case of dense stars, and a very hot environment as in the case of extremely bright stars. Therefore, the light of stars may be blue-shifted or red-shifted depending upon the atmosphere in which the hydrogen gas is present in the vicinity of those stars."

"Are these proven facts?"

"Gravitational effect is well proven, though the effect of high temperature is just an idea."

"Okay," he said, satisfied.

"But if it is assumed that hydrogen present in a star's atmosphere is not affected by any such conditions, still there is a chance of the light coming from stars being red-shifted or blue-shifted," I added further.

"You mean even if the energy difference between orbits remains unchanged?" He wanted more clarification.

"Yes."

"How?" he asked.

"There are two possibilities. One is due to the Doppler Effect, and the other is due to loss of energy during the travel of light waves through space."

"I have heard of the Doppler Effect," he interrupted, and added, "but that is related to sound."

"The Doppler Effect is not only related to sound, it can also be applied to other waves," I clarified.

"Can you explain this? I don't recall it fully," Gaurav requested.

"No problem. Imagine a man beating a drum with a constant rhythm of one beat per second."

"Okay."

"In one minute you will obviously count sixty beats, right?"

"Right."

"Further, imagine that the man is now moving away from you at a constant speed of ten meters per second while he maintains the pace of drum beating."

"Okay."

"Now how many beats will you hear in one minute?"

"Should be sixty."

"No. It will be fifty-eight." I corrected him.

"Why?"

"In sixty seconds he will reach a distance of 600 meters from you."

"Yes."

"The speed of sound is approximately 340 meters per second, so each beat will take more than one second to reach you when the drummer is at a distance of 600 meters from you."

"Right."

"In sixty seconds you will hear only fifty-eight beats because the fifty-ninth and sixtieth beats will be on their way. They will not reach you within the first minute since the time the drummer started going away from you. Thus, you will measure the frequency of the drum beating as fifty-eight beats per minute even when the drummer maintains the frequency as sixty beats per minute," I explained.

"Oh! Now I understand."

"This is known as the Doppler Effect. If you understand this effect, tell me how many beats you would count in a minute when he's standing 600 meters away from you?" I asked him to ensure that he had understood properly.

"Should be fifty eight beats per minute," Gaurav replied.

"No. That is wrong. When he is not moving away from you, every beat will take equal time to reach you, so you will again count sixty beats per minute, if he maintains his rhythm," I again corrected him.

"Then there should be sixty-two beats per minute when he is returning?" Gaurav reasoned.

"Correct, but only if he returns at the same speed as he had gone away from you." I added a condition.

"What will happen if he comes faster?"

"Then you shall count more than sixty-two beats."

"Okay. But, what is the use of all this?"

"Just by standing at one place and counting the number of beats per minute, you can tell whether the drummer is going away from you, standing there, or coming towards you; and you can even know whether he has increased or reduced his speed while traveling, provided he keeps the rhythm of beating the drum unchanged."

"That is an interesting deduction! Just by counting the number of beats, I can get so much information about the location and movement of the drummer." Gaurav was amused.

"Similarly, just by measuring the frequency of the light coming from a star, we can tell whether the star is going away, is stationary, or coming towards us. We can even calculate its speed from the variation in the measured frequency of light and the standard frequency of light."

"That is again a very interesting correlation!" Gaurav was amazed at this revelation.

"Yes, it is. There is one more possibility," I reminded, "You must have seen that if you throw a stone in a pond, ripples are formed."

"Yes."

"These ripples keep on becoming elongated as they travel outward."

"Yes."

"Thus, as the ripples progress outward their wavelength (*Distance between two consecutive crests or troughs of the wave*) increases in proportion to the distance traveled by them."

"Right."

"Now, you might remember that the product of wavelength and frequency of a wave is equal to speed of the wave."

"No, I don't remember," Gauarv confessed honestly.

"No problem. I just wanted to tell you that if the speed of a wave remains constant then its frequency and wavelength are inversely proportional to each other."

"Okay."

"That means the frequency of ripples measured by someone standing at a distance from the point of throwing a pebble will be less as compared to the frequency of ripples at the point of their origin," I explained.

"Got it," Gaurav admitted.

"Similarly, the frequency of light may also be reduced due to the travel of a light wave through space." I related the conclusion of the small experiment of pebble-throwing in the pond, with reduction of frequency of light coming from stars.

"And that will cause red-shift," Gaurav understood.

"Yes, and you may note one important point in this scheme, that the longer the distance traveled by the light, the greater the red-shift," I added.

"So, red-shift can be due to two reasons- one is the Doppler Effect and the other is the traveling of light through space."

Gaurav tried to sum up.

"The third one is the effect of high gravity near the stars. There may be more reasons due to different conditions of the environment of stars on the hydrogen present there," I reminded him.

"Okay," he agreed with me and then asked, "How is this red-shift related to the expansion of the universe?"

There was an interruption as the Captain of the airplane made an announcement about the remaining flight time and the weather in Mumbai. It would be an hour and a half before we landed in Mumbai.

The Big Bang and Lines of Space

Chapter 6
Expansion of the Universe

After the announcement in English, as well as Hindi, was over, I resumed telling him about the expansion of the universe, "In the beginning of twentieth century, astronomers got interested in measuring the red-shift and blue-shift of light coming from the stars, which could tell them something about the speed of the stars based on the Doppler Effect,"

"Okay."

"They had expected the light of some of the stars to be red-shifted and some as blue-shifted; implying that some could be going away from Earth, and some could be coming towards Earth, if the Doppler Effect is applied. But they found the light of the majority of the stars to be red-shifted."

"That meant that the majority of the stars were going away?" he asked in surprise. "If you consider only the Doppler Effect, then you can say 'yes,' but if you consider other reasons also for red-shift, then maybe 'no'," I suggested.

"Why?"

"Even if gravity, magnetism, and the brightness of a star are not taken into account, the red-shift should still account for two factors: the receding speed of the star and its distance from Earth." I reminded him.

"Yes."

"Since all the stars are at great distance from the Earth, the factor involving red-shift due to distance will always be positive, whereas the factor involving Doppler Effect may be positive or negative, depending on whether the star is going away or coming towards the Earth. Thus, a resultant of two factors have more probability of coming out positive. For that reason, they most likely found the light of a majority of the stars red-shifted."

"That seems reasonable. What happened next?"

"Then the astronomers found that the stars that are more distant have higher red-shift. It essentially implied that the red-shift factor due to distance was more pronounced. But, unfortunately, this factor was not considered at all," I informed him.

"Why?" he asked in surprise.

"Because it was believed that there is no medium in space and hence frequency of light cannot reduce due to its traveling through space."

"So, only Doppler Effect was left?" he asked.

"Yes. And based on that factor only, an astronomer named William Huggins calculated the expected speed of stars."

"Okay."

"He further tried to check whether the speed of stars found in this way was correct or not. He knew the distance of those stars from Earth. He plotted the distance of stars from Earth on one axis of a graph, and the speed of stars on the other axis, which was basically red-shift only. He found it to be almost a straight line, with only a few exceptions." I made a graph as shown in Figure 3.

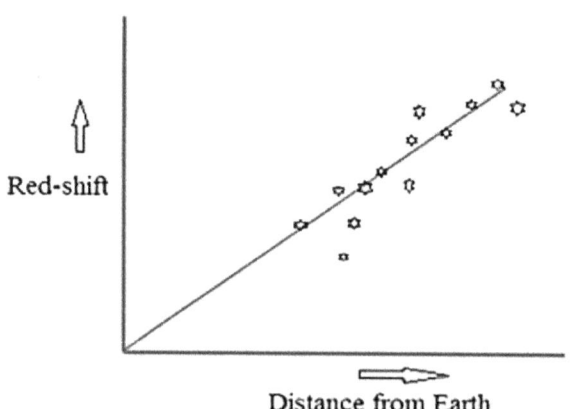

Figure 3
Graph of red-shift of light coming from stars and their distance from Earth

"Didn't it imply that there was a direct relation between the distance of a star from Earth and red-shift of the light coming from that star?" Gaurav reasoned as he was well aware of simple graphs used in engineering.

"Yes, that is correct. Although the graph essentially showed a direct relation between the distance of a star from Earth and red-shift of the light coming from that star, Huggins proposed that there is a direct relation between distance of a star from Earth and speed of that star by considering red-shift to be equivalent to the speed of the star, based solely on the Doppler Effect."

"Okay."

"He said that if a star is at twice the distance of another star from Earth, then that star is going away from Earth at twice the speed of the second star; the further a star is, faster it is going away."

"That is interesting."

"From this relation, the astronomers proposed that the universe must be expanding," I supplied.

"So, that is the explanation for expansion of the universe?" asked Gaurav.

"Yes. But you can see that this explanation is not totally accurate, because it assumes the red-shift to be solely due to the Doppler Effect," I cautioned him.

"Can I clear something for you?" asked the air hostess.

I picked up the empty glass from my tray, handed it to the air hostess, and folded the tray. Gaurav also did the same.

"Did you say Huggins, or Hubble?" he asked after the air hostess left, as he thought I had made a mistake in naming the astronomer.

"No, William Huggins was the astronomer to propose the relation between red-shift of the stars and their velocities using the Doppler Effect, whereas Hubble had discovered the first galaxy other than our own Milky Way. He also made a graph of the relation between distance of the galaxies and their velocities using the relation invented by Huggins," I clarified.

"But, we always hear the name of Hubble; and not any

other astronomer."

"Because Hubble got very famous due to his association with the Big Bang model. He inadvertently provided reasoning to the idea of the Big Bang proposed by a Belgian priest and famous physicist, George Lemaitre. He had proposed that the Universe must have originated from a big primeval atom. His idea was based on the Second Law of Thermodynamics that the entropy (*randomness*) of the universe is always increasing. Therefore, in the beginning of the universe, the entropy was minimum, and hence, all the universe must have been enclosed (or *well organized*) in a big primeval atom.

That was the first version of the Big-Bang. Hubble's premise that if the velocity of galaxies is related to their red-shift, then all the galaxies are receding from us, and since they are receding at a speed proportional to their distance, they must have been together at some time in the past, harmonized with Lemaitre's concept of the primeval atom."

"That meant that all the galaxies started going away from one another due to some explosion, and that explosion was called the Big Bang," Gaurav reasoned.

"Yes, that is the basis of the Big Bang theory, but Hubble had said that 'if' the velocity of galaxies is related to the red-shift of their light." I pointed out the significance of 'if' in Hubble's statement.

"That means he was not certain of this relation?" he asked.

"Yes, he was not certain about it because that relation was a fairly new concept proposed by Huggins. Though it was still not proven beyond doubt, Hubble's idea got associated with the Big Bang theory and it became so popular that 'if' was dropped from this statement, and it was assumed that the relation between velocity of galaxies and their red-shift was true."

"So you're saying that the basis of the Big Bang theory may not be correct if the use of the Doppler Effect is erroneous in calculating the velocity of stars from red-shift," Gaurav tried to understand my logic.

"Yes, and, in fact, many quasars were found in the universe which had very high value of red-shift. If we use the Doppler

Effect only to find the amount of speed that quasars required to have that kind of red-shift, that will show them receding at speeds comparable to the speed of light, and far far away from us, but their actual distance is found to be much less."

"What are quasars?"

"Quasars are very bright stars and they send strong radio signals. They are believed to rotate very fast."

"How was their actual distance found?" he asked.

"Similar to the way astronomers measure the distance of a star. There are some stars that have variable brightness, they are called variable stars. Their brightness varies in a fixed pattern. It may keep on reducing for many days and then suddenly increase in a few hours, or may take a longer time, but the pattern is fixed and is repeated for every cycle for a variation of brightness. Astronomers note down the brightness variation period of these stars. Then they calculate their true brightness from a relation between the true brightness of the star and its variation period. True brightness is then compared with apparent brightness of the star to give the distance of the star from Earth, because the brightness of the star reduces by the square of the distance. This technique gives the distance of stars and galaxies."

"Okay."

"The distance given by red-shift is not a correct measure in the case of quasars. This fact also creates qualms in the expansion theory of the universe." I pointed out.

"Are you saying that the expansion of the universe is an erroneous notion and, hence, there was no Big Bang?" Gaurav asked.

"Yes. The Big Bang theory was based on a doubtful idea that the universe is expanding, so it required definite evidence that could prove this theory, but there is no clear-cut proof, as yet."

"How can evidence be found for this kind of theory? It happened billions of years back. How can any evidence survive that long? I think you just have to believe it without the evidence," Gaurav suggested.

"In science, no theory is valid if it does not have evidence."

"Then why is this theory so popular if it is not valid?" he asked.

"Because claims have been made for finding evidence of this theory."

"What claims?"

"Two main pieces of evidence were claimed in support of this theory about fifty years back. One was based on radiation and the other was on the abundance of elements in the universe," I informed him.

"Was the theory proved after these claims?" he asked.

"Actually those arguments also were not infallible. There were problems with those, too."

"What?"

"I shall first tell you about the claim based on radiation."

"Okay."

Chapter 7
C.M.B.R

"The Big Bang theory proposes that all the stars and galaxies were clumped together in the beginning of the universe. Then the particles in that small cluster must have been very dense, and their temperature must have been very high due to numerous collisions between the particles at close range. Thus, the temperature of the universe at the beginning must have been millions of degrees centigrade. At a temperature above 2700 degrees centigrade, atoms of all elements become so energetic and excited that the electrons do not remain bound to the nucleus. The nucleus and electrons are free to move around without forming atoms. In this stage of the environment, light would not have been able to travel straight without colliding with either positive ions or negative electrons, and thus, it would have followed a zig-zag path similar to the situation of fog in the winter season."

"Hmm," Gaurav uttered, and indicated for me to go on.

"As our sight gets limited in foggy condition, similarly, there must have been a condition of darkness everywhere at that time. This condition was calculated to have existed until 300,000 years from the beginning of the hypothetical Big Bang, until the temperatures cooled down to below 2700 degrees centigrade to allow the atom formation."

"That is an interesting thought," Gaurav commented.

"When the atoms formed, light was free to travel straight for a long distance for the first time. The fog got cleared and the universe lit up for the first time. Light released at that time is believed to be traveling in the universe even today. This light, if found, could prove the theory of the Big Bang"

"Oh my God! What an idea!" Gaurav was amused and asked, "How can we find this light?"

"It was accidentally found in 1963 when two radio operators were trying to fine tune their radios."

"Accidentally?"

"Yes, in fact, they were looking for something else. They were given the job to eliminate the disturbance from the radiation signal which interfered with the transmission of sound by radio. They checked all their equipment many times over for any loose connections, but they could not remove all the interference. Then they thought that the pigeons sitting on their antenna were causing this trouble. So, they caught all the pigeons and removed them from the area."

"The pigeons used to sit on the antenna of our television also. I remember seeing the high antennas of televisions perched on almost every roof in the city about twenty years back, when I was a child," Gaurav recalled and then asked excitedly, "What happened then?"

"You know, pigeons remember their place of abode; they were used as couriers in olden days because of this ability."

"Yes, that's right." He nodded in agreement.

"So, the next day, they found the same disturbance in their signal when they switched on the radio. All the pigeons had returned."

"Hee hee hee!" he chuckled quietly.

"They caught all the pigeons again and hired a man to get rid of them. He strangulated all the pigeons one by one in exchange for some money."

"Oh no! So cruel!" said Gaurav with revulsion.

"Then they switched on their radio and found that the disturbance was still there."

"So, all the pigeons lost their lives for nothing."

"Yes. They had to search elsewhere for the cause of the disturbance. They tried shifting the antenna in other directions and hoped for any change in the disturbance, but they found that the disturbance was the same in every direction."

"Was this related to light from the early universe as you had mentioned?" Gaurav asked as he got the link now.

"Yes, this radiation had the same wavelength as was

expected from the light released from the early universe, but the problem was that these radio operators did not know anything about the conditions of the early universe, as they were not cosmologists. Knowledge about the remnants of light coming from the initial conditions of the universe was limited to the cosmologists only."

"You had told me about finding the light from the early universe, but this was some radiation." Gaurav pointed out the difference very intelligently.

"That light was calculated to have converted into microwave radiation in the billions of years that had passed since then, due to the expansion of the universe by more than 1000 times," I clarified.

"But you said there was no expansion."

"This is according to the scientists who believed in the Big Bang and expansion of the universe," I reminded him.

"Okay. Then what happened?" Gaurav wanted to know more as the story was becoming more and more interesting.

"Coincidentally, they met one cosmologist and shared their finding with him. He was, at that time, uninterested in their story, but later, he realized the value of their findings and wrote a letter to them to divulge the fortunate finding they had stumbled upon, which later caused them to win the Nobel Prize in physics," I informed him.

"Strange way of finding this radiation, but it must have given a big boost to the Big Bang theory," commented Gaurav.

"Yes, it gave a much needed boost to the Big Bang theory," I agreed.

"Still, you say that this claim of evidence of the Big Bang was doubtful?" he asked with a playful smile as if I was caught in my own woven net.

"Yes, I still say that the Big Bang is doubtful," I replied serenely.

"Why?" he asked as he sensed that there was something more to it; the story had not ended.

"There was a problem related with this finding. This radiation coming from all directions was observed to be homogeneous," I furnished.

"What is the problem with that?"

"Homogeneous radiation means the same temperature of the universe in all directions."

"I don't get it."

"It meant that the whole universe was at a uniform temperature when the light was released for first time."

"So what?"

"The temperature could not have become uniform, as there was not enough time to equalize after the Big-Bang."

"Wasn't it 300,000 years after the Big Bang? That is a long time to equalize." He remembered the figure which I had given him for release of light.

"But according to the theory, the universe kept on expanding in those 300,000 years. The diametrically opposite ends of the universe, from where this radiation was coming, were measured to be ninety times more distant than the light could have traveled from the beginning of the universe. Since nothing can travel faster than light, the temperature of the regions lying beyond the reach of light could not have equalized. It meant either the date calculated for the Big Bang event was wrong or there was some other explanation. Since the date of the Big Bang was fixed according to the receding speed of stars and galaxies, it could not be pushed back. There had to be some other reason. So, the finding of Cosmological Microwave Background Radiation, as this radiation was called, created another problem, which was given the name 'horizon problem.'"

"Oh!"

"Uniform temperature of this radiation also implied that the universe should be uniform in structure, whereas we can physically observe that there are a wide number of galaxies and clusters of galaxies where the matter is concentrated, and there are large intergalactic spaces where there is no matter. Thus, the uniform radiation coming from the hypothetical scattering surface of the Big Bang did not furnish reliable proof."

"That radiation is still present or not?" he asked.

"It is present even today," I agreed.

"Then from where is that radiation coming if not from the Big Bang?" he asked.

"Good question." I appreciated him. It showed that he had genuinely listened to my lecture.

"Your question is valid. What is the source of radiation if it was not from the Big Bang?" I repeated his question, and then replied, "There are so many radiations in the whole universe, there has to be an aggregate of all that which will be spread out equally in all directions, and will have one wavelength everywhere. Consider it something like sea level. In the sea, there are waves of different heights but when they are averaged out, they will form one level; that will be sea level. Similarly there has to be one level of aggregate radiation emanating from all directions equally," I explained.

"Seems logical," remarked Gaurav and added, "Then there will not be any horizon problem either."

"Right."

"You had said there was one other proof."

"Yes, that was the abundance of elements." I started getting up to go to the washroom and said, "I'll be back soon." He understood.

The Big Bang and Lines of Space

Chapter 8
Abundance of Elements

More than an hour had passed as I talked with Gaurav about the Big Bang. I had gotten a little tired trying to recollect various facts about the Big Bang theory, which I had learned a few years back. Although they were not part of any of my educational curriculum, I had learned about them due to my special interest in the subject. I never thought that I would be called upon to discuss them while traveling, but I liked the conversation. I had already given Gaurav a great deal of information about the Big Bang. When I came back, he was waiting for me to resume the story.

"Scientists calculated the abundance of light elements in the universe, in comparison to hydrogen, that could be created if the Big Bang really occurred. They found that helium would be about 25%, deuterium one hundred thousandth part, and lithium about one part in ten billion. This figure matched with the actual abundance of elements found in the universe." I told Gaurav.

"Wow! This is good proof!" Gaurav said excitedly.

"Don't get so excited: there is a catch in this, too."

"What?"

"For making this calculation, physicists assumed certain matter density – the ratio of the matter to the volume of the entire universe. Now, they had to verify whether the matter density used for that calculation was correct or not."

"Oh!"

"The scientists calculated the mass of all observable matter in our galaxy, and observed that the matter density was only half of what was required for proof of the Big Bang."

"Couldn't there be more matter in our galaxy than what is observable?" he asked.

"No, all the matter in our galaxy was carefully calculated."

"Then what happened?"

"It was thought that there must be some matter that is not visible, and it was called 'dark matter.'"

"So, dark matter was invented to corroborate the Big Bang?"

"Yes, but the problem was not yet sorted out because validation of dark matter was missing. So, another method was devised for calculating total matter in the universe."

"Which method?"

"Scientists used the gravity and speed of the outermost parts of our galaxy to calculate the quantity of matter inside the galaxy."

"How could matter be found from that?"

"Matter is related to gravity, and the speed of the outermost parts of galaxy are also related to gravity through centrifugal force. So, they calculated the amount of matter required in the universe to speed up the outermost parts of our galaxy. Since speed was measured accurately by them, they expected this calculation to give an exact amount of matter density in the galaxy and settle the dispute."

"What did they find?" asked Gaurav.

"It was found that there should be ten times more matter than was observable."

"Oh! It overshot the requirement for the Big Bang even?"

"Yes. It implied that there was much more dark matter in the universe, if it existed. But this finding definitely proved that the abundance of elements was not watertight proof of the Big Bang theory."

"So, both pieces of evidence for the Big Bang theory were not dependable?"

"Yes."

"Then what?" He now began to doubt the Big Bang model and commented, "Really, the Big Bang model had plenty of problems. I never knew this."

"Yes, it had. Besides the dark matter, the physicists also had to invent a new term – 'dark energy'."

"Why?"

"They calculated the energy required for the Big Bang model by using Einstein's Theory of Relativity and found it to be one hundred times the total energy of the universe, calculated by other means."

"Oh! One more problem!" sighed Gaurav.

"Yes, and then to reconcile this finding of the Big Bang theory, physicists called the missing energy 'dark energy,' and suggested that this energy must be present in some other particles that are not yet discovered. Instead of discarding the Big Bang model, which gave such erroneous values for matter and energy, they opted to find new particles."

Gaurav was amazed with these revelations. I could judge from his expressions that he was now fully convinced of the ineffectiveness of the Big Bang theory. I continued further, "Due to these problems, cosmologists requested help from particle physicists to find new particles which could account for dark matter and dark energy."

"Did they find them?"

"I think we should take on that discussion later, our plane shall be landing soon."

"Just this one before landing," requested Gaurav.

"They suggested *nothing*."

"They could not find it?"

"No, they found it and suggested it as I told you."

"But you said they suggested nothing."

"Yes, they suggested *nothing*, which they claimed could convert into particles and antiparticles. It is like having a zero, and then saying that you have +1 as well as -1. You can use either of them, whichever you like, and leave the rest. So, *nothing* can convert into particles and antiparticles. You use the particles and leave the anti-particles which are of no use for you. This way you can have the particles required for reconciling the Big Bang theory. See, they did not disappoint the cosmologists," I said with a smirk.

"You are certainly joking now, it is not possible." Gaurav sniggered.

"Anything is possible in modern physics," I said with a smirk.

"I can't believe this bizarre hypothesis," Gaurav expressed his feelings.

"There are more like this," I said and reminded him about a statement I had made earlier, "I had told you about the problem of homogeneous radiation coming from all directions of the universe that made it difficult to explain the existence of the super-clusters of the galaxies."

"Yes." He recalled.

"Physicists offered a solution to that also."

"What was that?"

"Inflation theory." Though I had wanted to wind up the session, I realized that wasn't going to happen as soon as I uttered those two words.

Chapter 9
Inflation Theory

We still had about half an hour before the plane would land, so I started talking about the Inflation theory. "In the first second of formation of the universe, immediately after the Big Bang, the universe expanded to double its size many times over. Each time it took only one billion-billion-billion-billionth of a second to double in size. This process went on for about one hundred times and caused the universe to expand at an unimaginable rate."

Gaurav looked at me in question.

"If you make double of one, you get two; and if you again double it, you get four. Similarly, if you keep on doubling it repeatedly, you will reach a figure of more than one million after doubling it twenty times, and more than one million-billion after doubling it 50 times. If we assume the radius of the universe as one micro-meter (one millionth of a meter) at the start of the inflation, then it would become the size of one million kilometer in less than a billion-billion-billionth part of one second."

"Is the universe expanding so fast?"

"No, the physicists hypothesized that the universe must have been inflating in this fashion for the first fraction of a second. Only then could it take care of the problem of homogeneous radiation. Thereafter, it was believed to have slowed down due to the effect of gravity."

"I am sure this model must also be having some other problem, I am not going to put my money on the Big Bang now," Gaurav sneered.

"You are right. This inflation was billions of times faster than the speed of light."

"Oh! Now I get why you have doubts about this theory,

because nothing can move faster than light, and this inflation was faster than light," deduced Gaurav.

"No. For that, physicists gave another reason–although nothing can travel faster than light through space, space itself can travel faster than light."

"What do you mean?"

"When you are walking inside a moving train, there can be a limit to your walking speed, but you cannot put a limit to the speed of the train. You cannot say that train should run only at lesser speed than your walking speed. Similarly nothing can move faster than light through space, but that does not restrict space itself to expand at a speed more than the speed of light," I tried to explain.

"Okay. If the speed of expansion was not the problem, then what was the problem with Inflation?"

"The Inflation Model is based on such assumptions that cannot be proved; you just have to assume some initial conditions, which are unbelievable, such as magnetic monopoles and negative gravity. These things don't exist, so it is very difficult to prove or disprove this theory."

"What are magnetic monopoles and negative gravity?" Gaurav wanted to know further.

"I shall reply to that a bit later," I suggested as we heard the announcement from the Captain of the plane, thanking all guests and instructing his crew to get ready for landing.

"Okay," agreed Gaurav, and started looking outside where the shoreline of Mumbai was coming into view.

Chapter 10
God Particle

After we reached the baggage claim area, Gaurav asked, "Sir, finally, has the Big Bang been proven or not?"

"No. Not yet."

"The Inflation model was also rejected?"

"No. I told you that such models can neither be proved nor disproved."

"If the universe was not created according to the Big Bang theory, what is the other possibility?"

"There is a possibility that everything has been created from space itself."

"What! You said…everything in the universe is made from space itself. Right?" he thought he had heard wrong and added, "This is not possible; how can space get converted into particles of matter?"

"Remember, I told you about zero becoming -1 and +1?" I asked him.

"Yes."

"That was not a joke. Physicists are serious about that."

"I can't believe it!" He was amazed.

"Do you think that empty space is really empty?" I asked him in order to start explaining the concept.

"Yes, it is," replied Gaurav as we retrieved our bags.

"You know, some recent discoveries indicate that there is something in all that empty space," I told him as we started walking towards the exit after putting the bags onto trolleys.

"Is it so?"

"Have you heard of the discovery of the 'God Particle'?"

"Yes, a little bit."

"Peter Higgs and Ernest Francois have received the Nobel prize for discovery of the God Particle," I told him as I kept looking for someone holding a placard with our names.

"Does it mean that a part of God has been discovered?"

"The God Particle has nothing to do with the existence of God; it is just a name given to a particle that was believed to exist by Professor Peter Higgs and Ernest Francois in independent studies in the 1960s."

"Why was this particle named the God Particle if it has nothing to do with God?" Gaurav was more interested in knowing about the God particle than worrying about our contact at the airport. I was afraid that if nobody came, then we might have to search for a hotel ourselves. There was nobody with a placard with our names on it.

I took out my mobile phone and switched it on. It was taking its own sweet time to get connected to the new network, so I told Gaurav, who was waiting for an answer from me, "This particle has been the most sought after particle by all the scientists in the world. It remained invincible for a very long time. Physicists had doubts whether this particle really existed or not, so they called it the God Particle and the name stuck with it."

"Oh!" exclaimed Gaurav and, at the same time, my mobile rang. I was not expecting a call almost immediately after landing in Mumbai. It was an unknown number but I answered the call. The man on the other end was our contact. He had gotten delayed in reaching the airport due to traffic, but wanted to let me know he would be at the airport in five minutes. I informed Gaurav about that, but he seemed more interested in the story of the creation of the universe, and I could make out from his facial expressions that he wanted me to continue the story.

We had to wait for our contact anyway, so I continued, "It is actually named the Higgs Particle, after Peter Higgs, who proposed the existence of this particle. The God Particle name became popular for this particle. Although it was wrong to call it that, but in order to generate funds from the various governments to conduct experiments to find this particle, marketing experts of the scientific community stuck to this name."

"It was a marketing stunt!" Gaurav was surprised.

"Yes, you could say so. It was even called 'goddam particle' due to the frustration of scientists to find it."

"Ha ha... goddam particle!"

"But that name did not become prevalent. The discovery of this particle was announced by scientists at Linear Hadron Collider, Geneva in July 2012. After that, the proposer of Higgs particle was awarded the Nobel Prize for Physics, as I told you. In his theory, he also proposed the Higgs Field pervading in empty space throughout the universe, from which the Higgs Particles erupt."

"What is the Higgs Field?"

"It is a field like the Gravitational field. The Gravitational Field is supposed to be spread in space around the sun, the stars, and Earth, etc. But the Higgs Field was proposed to be spread everywhere in space, even in a vacuum, irrespective of the presence of any matter such as the stars."

"You mean that the Higgs Particle can erupt from a vacuum also?"

"Yes, and it implies that empty space is not really empty."

"This is interesting; I had always thought that space is really empty."

"So, if space has something in it, from which Higgs Particles can erupt, then there is a possibility that other particles can also erupt from space. Thus, the stars could have been created from space itself because all stars are created from particles."

"Wow!"

"So far, physicists have only found the Higgs Particle; they have yet to find the exact mechanism of transformation of space into matter. But they have a theory."

"What is that?"

"At very high energy, space can convert into matter and antimatter, as I told you earlier, and particles can erupt from nothing."

"What is antimatter?"

"Scientists say that when the matter was created from space, antimatter was also created, similar to our example of creation of +1 and -1 from zero."

"Then where is that antimatter?"

"It is believed that there may be a universe solely made of antimatter, but it will be exactly like our universe, as anybody living in that universe will think our universe is made from antimatter and consider theirs to be made of matter."

"This seems to be speculation only," Gaurav objected.

"Yes, science is full of such speculations, like multi-world and multi-dimensional worlds."

"This guesswork appears as if it is from science fiction books," Gaurav commented.

"It seems our man is here," I said as I saw a Nepali boy exiting a van and holding a placard with our names.

Chapter 11
At the Guest House

In the guest house arranged for our stay, there were three bedrooms with attached bathrooms, one common room with a dining table, a couch with center table, and a 42-inch LED television. It also had a small kitchen with a gas stove for cooking, a refrigerator, and was equipped with a few utensils for cooking and serving food. The attendant of the guest house was another Nepali boy named Sunder, who showed us to our rooms.

The third room was occupied by an electrical engineer from our company, Mr. Narender Singh. Although we had never met previously, working for one company was enough for us to greet each other warmly. After a brief introduction, we decided to refresh ourselves and ordered Sunder to make tea for us.

I arranged my belongings in my room and changed into casual gear before sauntering back to the common room. Sunder had made tea and set it out for us on the dining table. I poured tea into a cup from the flask and parked myself on the settee opposite Mr. Singh, who was seated in another chair, watching a news channel on the television. He had a well toned body, and at just under six feet, would be considered tall by Indian standards, where the average height of a man is five and a half feet. He seemed very energetic and garrulous by the way he had talked to me as we came in. Had his receding hair line not given away his age, I would have guessed him to be much younger than in his late forties. He looked at me and smiled as I sat in front of him.

"*Batti Sahib*, how long have you been working with this company?" I asked him, addressing him by the commonly used title for Electrical Officers on board Indian manned ships. (*Batti Sahib* is the Hindi translation for Electrical Officer.) I couldn't call him by his name, as he seemed about the same age as I, or older. It is rude to call elders by their first name, according to Indian customs. He also couldn't call me by my first name because I held a senior position. The best option for us was to address each other by titles because each of us was expected to give respect to the other. These are some unwritten rules of Indian society which you learn as you grow.

"It is ten years now," he replied, then asked, "What about you, sir?"

"This is my nineteenth year," I replied.

"Oh! That is an achievement!"

"I had no reason to leave this company, so I stayed here," I gave as my reasoning.

Gaurav also came in, poured his tea from the flask, and took a seat next to Mr. Singh. After having a sip from the cup, he asked Mr. Singh, "*Batti Sahib*, any idea how we'll get to the Institute?"

"A van will be here at eight o' clock in the morning to take us," replied Mr. Singh.

"Thanks. Do we have Wi-Fi here?" he asked.

"Yes, Sunder has the password," Mr. Singh informed him. He had reached the guest house one hour before us and had made the inquiries from the attendant.

"Thanks again," he said to *Batti Sahib* and then addressed me. "Sir, it was nice to learn about the Big Bang during the flight."

"Thanks."

"I want to know more about that."

"What are you talking about?" asked Mr. Singh as he looked toward Gaurav.

Gaurav recounted the substance of our dialogue during the flight. Mr. Singh listened to him. Gaurav did an excellent job of explaining the ambiguities in the Big Bang theory. I only had to chip in a few times to clarify some points.

In the meantime, Sunder had procured dinner and asked us if it would be okay for him to start serving. He had brought pre-cooked food for us. In fact, the owner of the guest house had made an arrangement to supply cooked food for all the guests from one central location. He operated a network of eight guest houses in close vicinity, and found it cheaper to cook food at one place and deliver it in steel boxes to various guest houses. The attendant was left with only the job of serving it and preparing hot *chapattis* (Indian bread). Sunder told us about this arrangement as soon as Mr. Singh asked him whether he had prepared the food himself or brought it from outside.

We chatted with Sunder for some time while he arranged dishes on the dining table. When he left to make *chapattis*, we moved to the dining table, seated ourselves, and waited for the *chapattis,* which are generally served hot. During dinner, Gaurav kept narrating the details of our previous conversation to Mr. Singh. After we finished dinner, he asked me, "What are magnetic monopoles and negative gravity?" He remembered where I had left off telling him about the Big Bang.

"They are what their names suggest," I told him.

"Can you explain them?" he requested.

"Take the case of an elevator. It is pulled down by gravity, but for going up, it requires power. If it can go up without any power as if something is pulling it up, just as the gravity pulls it down, that force will be called negative gravity," I replied.

"What about magnetic monopoles?" he asked further.

"Every magnet has two poles; one north and one south. If we cut that magnet into two pieces, we will get two magnets having two poles each. If you can have a magnet with only one pole, that will be called magnetic monopoles. But you cannot have a magnet with a single pole, irrespective of the number of times you cut that magnet." I explained about magnetic monopoles as well as suggested the improbability of finding them.

"That is right," agreed Mr. Singh.

"But scientists believe that there existed magnetic monopoles in the very early universe. They are trying to find them as evidence of the Inflation theory." I added.

"Oh! If they find them, will that prove the Big Bang theory?" Gaurav asked as he saw some hope for that theory.

"I can't say. If magnetic monopoles are found, they may cause some other problem." I gave away my suspicions.

"And what will the negative gravity prove?" asked Mr. Singh.

"If there is negative gravity, then any time something fell apart all its fragments would keep on going away from each other, just like an explosion. Scientists are depending upon negative gravity to explain the unbelievably high speed of inflation in the first second of the creation of the universe. As per Inflation theory, in less than a billionth of a billionth of a billionth of the first second of the creation of the universe, it expanded to billions of miles due to negative-gravity. After reaching a certain distance, negative gravity was converted into gravity, and the expansion of the universe started slowing down. If negative gravity is found, that may prove this theory," I explained.

"That is fascinating!" remarked Mr. Singh.

"Yes, it is a fascinating theory, but it cannot be proven until the existence of magnetic monopoles and negative gravity are established, and that seems impossible. To me, it appears like Ptolemy' epicycles," I uttered without realizing that they would not be aware of that either.

"What are Ptolemy's epicycles?" was obviously the next question for me from Gaurav.

Chapter 12
Ptolemy's Epicycles

"Ptolemy was an astronomer who lived in the 2nd century AD in Alexandria. He had devised a way of explaining the motion of the planets around the Earth using epicycles, which were believed to be real for nearly fourteen centuries until Galileo proved their non existence."

"But planets revolve around the Sun," Gaurav interrupted.

"Scientists of that time thought that the Earth was the center of the universe and all the planets and the Sun revolved around it. According to them, most of the planets followed this rule except five of them – Mercury, Venus, Mars, Jupiter, and Saturn, because they observed that the paths of these planets were erratic. These five planets generally moved around Earth in a circle, but sometimes they seemed to stop and reverse their direction."

"Why?"

"Because they were actually moving around the Sun, not the Earth."

"I don't understand this point." Mr. Singh wanted further clarification.

"When you are traveling in a train and looking out of the window, the trees on the side seem to be moving backward against the background of some distant, fixed object."

"Yes, it happens."

"We are on Earth, which is moving, and when we see any planet against the background of fixed stars, it will appear to move backward if it is stationary, similar to the case of trees just mentioned," I explained.

"Okay."

"If the train is not moving, then the trees will not appear to be moving backward," I said further.

"Yes."

"The astronomers at that time were under the impression that the Earth is not moving, so when they observed the planets moving backwards, they tried to provide a reason for the backward motion of these planets while keeping the Earth stationary at the center of solar system."

"But the planets are not fixed like trees, are they?" expressed Gaurav.

"You are right; these planets move in the same direction as Earth moves, but if you go back to our example of the train and the trees, and replace the trees with slow moving vehicles such as a bullock cart etc., what will you observe?" I changed the situation in order to explain the movement of planets.

"In that case also, the bullock cart should appear to be going backward," replied Mr. Singh.

"Right. And now you replace the bullock cart with a car cruising at a speed more than the speed of the train, what will you observe?" I changed the scenario again.

"Then, the car should appear to move in a forward direction," reasoned Gaurav this time.

"Right. Thus your observation from the train depends on the difference of the speed of the train and the intermediate object, between you and the distant fixed object. If the object is moving faster than the train, it will appear to move forward, and if it is moving slower than the train, it will appear to move backward," I clarified.

"Okay." Mr. Singh and Gaurav understood.

"But, when the train is not moving, will you see the tree or the bullock cart moving backward?" I asked them.

"No," replied Gaurav confidently.

"Similarly scientists of that time were troubled. Why did the planets appear to move backward when the Earth is stationary?"

"They could have simply said that the planets sometime move forward and sometimes backward; what was the problem with that?" asked Mr. Singh.

"Nothing can change speed on its own without the

application of a force, so planets cannot keep changing their speed abruptly while traveling in space. There had to be some other reason," I clarified.

"What did they find?"

"Ptolemy thought that while these planets revolved around the Earth, they also revolved around in smaller circles, which he called epicycles."

"How does the epicycle explain the backward motion of the planets?"

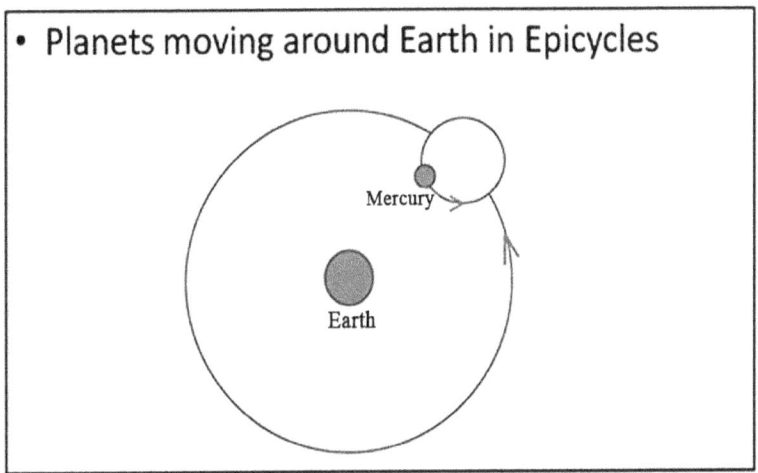

Figure 4
Ptolemy's Epicycles depicting backward motion of planets

"I'll show you a simple diagram." I took out my pen and drew a diagram on paper.

"Ptolemy imagined Earth to be at the center, and a planet – let's say Mercury, for example, is revolving around it in a circle as shown by the arrow in the bigger circle of the figure. Besides rotating in the bigger circle, he proposed that the planet also rotated in a smaller circle in the direction shown by another arrow mark. When the reverse speed of the smaller cycle (epicycle) would exceed the forward speed of the main cycle, the planet would appear to move backward," I explained with the help of diagram.

"So, by adding an epicycle to an orbit of a planet, Ptolemy gave a reason for the erratic motion of these planets," Gaurav understood.

"Yes."

"But why did you say that the Inflationary theory of the Big Bang is like Ptolemy's epicycle?" He remembered the context in which I had commented about Ptolemy's epicycle.

"When the basic idea of all planets revolving around the Earth was incorrect, then adding an epicycle to the orbit that just didn't exist was a tendency to keep sticking to an erroneous idea in an effort to demonstrate that the basic design was accurate, but required some minor modification. I can see a similar tendency in the Inflationary model just to save the erroneous basic model of the Big Bang." I gave my viewpoint.

"Now I understand your point," Mr. Singh agreed and then asked me, "Sir, how do you know so much about the Big Bang?"

"I read about it," I gave a casual reply.

"No, by just casual reading you cannot give the detailed facts and figures as you have given without referring to any book," he reasoned as he was not convinced with my answer.

"I have written a book on this subject," I revealed. Then both of them wanted to know about the book. The conversation took a different turn, and went on until we bade goodnight to one another.

PART II
ETHER

Chapter 13
Ether

The next day at the Institute, we learned about the new software used for controlling engines. The pick-up van was waiting for us when we finished our schooling for the day. As soon as the van pulled out of the Institute, it got stuck in the evening traffic. The van was inching its way slowly through the traffic when my colleagues started having a conversation with the driver.

The driver told them he was an experienced male nurse working as an assistant to a doctor. He had come to Mumbai just ten days ago to fulfill his dream of living here. Mumbai is famous for the film industry, and people from all over India desire to visit this city. This driver had left his job in a clinic and came here merely to fulfill his desire.

Driving was the first job he could lay his hands on after landing in Mumbai. His story was proof of how people from small towns got attracted to Mumbai. I marveled whether Mumbai had some attractive power similar to the gravitational attraction of Earth. This thought was ridiculous as the attraction towards a city depends on a person's own thinking and behavior. If the perception of a person cannot be used to define any physical law such as gravitational attraction, then why is the perception of an observer important in the Theory of Relativity?

I was half listening to the conversation of my colleagues with the driver while I contemplated on the Theory of

Relativity. My attention diverted unexpectedly at the sight of the people that were living on the roadside. The women were doing the laundry with the water pulled out from an opening in the pavement. There was a sort of tank arrangement under the pavement for the collection of rain water from the over-bridge. This water was being used for washing.

This showed the lack of even proper water facilities for the poor people living on the embankment of roads. There was obviously no space for proper sanitation, or laundry work, in their makeshift huts, made along the road with bamboo sticks and tarpaulin. I was distressed to see the plight of those people. Although I felt sorry for them, I had to admit that they looked happy; there seemed no complaint on their faces.

There was a small temple there also. It had no space for devotees to sit inside as there was just sufficient space for installing only a few idols. The people had to pay their obeisance by standing outside the temple. Of course, if the people did not have a place to stay, their God also had to be content with small space, whatever they could afford. There was a flag on top of the temple which indicated that the deity inside the temple must be Lord Hanuman (*The Monkey god with immense strength and resolute devotion to his master Lord Rama*).

I marveled that even though the people lived in such unsanitary conditions, their belief in God was not diminished. Scientists all over the world try to find the God Particle through expensive scientific experiments, whereas these poor people find God within themselves. I reckoned that all these people must have also come to Mumbai in search of their life's dream, but were not so lucky to find good jobs. Chances were high that this little roadside community would be the best any of them would ever have in life.

The van moved slowly in the heavy traffic; the sight of huts gave way to shops, other business establishments, and to some high rise buildings. I wondered how much disparity existed in that city as far as the status of people was concerned. There were ultra rich as well as very poor; still the city fascinated everyone who came there.

I was lost in my thoughts when I heard Gaurav. "Sir, *Batti Sahib* is asking you something."

"Sorry." I got startled as I came out of my rumination and apologized for not hearing what Mr. Singh had asked me.

"Sir, yesterday, you told us that the basics of the Big Bang theory are not firm, so the theory is probably incorrect." Mr. Singh made a statement instead of asking something.

"Yes," I agreed.

"According to you, the basic assumption that the red-shift of light is due to receding velocity of stars may not be correct."

"Yes." I had no idea where he was leading me.

"Then what can be the reason for the red-shift of light coming from stars and galaxies if that is not the reason?" He finally asked his question.

"Oh!" I sighed as I realized his issue. I looked around and asked him, "Do you want me to explain this here?" I indicated that we were in a van, and I did not consider it a good idea to have a discussion on science in a moving van.

"It is okay. The van is hardly moving, and we can hear you clearly. It will be a good way to pass time as well as gain knowledge." Gaurav supported Mr. Singh's request.

"But I had already explained to you." I looked at Gaurav.

"He tried to explain to me, but I couldn't get this point," Mr. Singh replied in defense of Gaurav.

"If you want to hear it, then I have no problem. I was slightly apprehensive that you might find some distraction on the road and then change the subject suddenly," I said with a smirk.

"Ho! ho! There is hardly anything beautiful around." Mr. Singh understood my remark and assured me, "Otherwise I wouldn't have asked you about red-shift now."

"Red-shift can be due to many reasons. One is obviously as used in the Big Bang theory."

"But you said that is incorrect."

"No, I didn't say that is totally incorrect...but this is not the only reason."

"Then?"

"It can also be due to the distance of stars, due to high gravity, due to high atmospheric pressure on the stars, due to rotation of the stars, due to extremely high temperature and brightness of the stars, or the combination of all these reasons."

"So many?"

"Yes. But most probable is the case of the distance from the stars."

"Hmm," Gaurav grumbled as he remembered our earlier discussion during the flight.

"Light is a wave and its energy is proportional to its frequency. When any wave travels through a medium, it loses energy. Its frequency may drop in proportion to the distance covered by the wave. This means that the frequency of light coming from stars may be reduced due to their distance from Earth," I explained.

"This had seemed to be a logical explanation to me also, as I told you," commented Gaurav looking toward Mr. Singh.

"Then why was this reasoning not used?" asked Mr. Singh.

"There was a problem related with this interpretation. It was proved by an experiment in 1887, conducted by two scientists, Michelson and Morley, that there was no medium of space, so light could not lose energy when passing through space. It had to reach the Earth in the same state as it left the star."

"Oh! Then this reason is not valid?" asked Mr. Singh.

"This is what was believed at the end of nineteenth century, but there are doubts about discarding the existence of the medium of space. To understand this properly, you shall have to go one and half centuries back in the history of science. In the 1860s, when British physicist James Clerk Maxwell had proven that light is an electromagnetic wave, he had assumed that it traveled on a medium of space."

"Why did he make this assumption?" asked Mr. Singh.

"You see, when we are talking, sound is traveling between us. Air is acting as a medium for sound waves to travel. If air is removed from here, we will not be able to hear each other because sound will not be able to travel. Similarly, when it was established that light is an electromagnetic wave, Maxwell thought that it must be traveling through a medium. Since light traveled in outer space where there is no air, he assumed that there should be some kind of medium even in outer space which helped light waves travel."

"Perfect. This is good logic for the medium of space to exist," Gaurav agreed.

"It was named Luminiferrous Ether because it helped the traveling of light. It could also simply be called Ether."

"Then what happened?" Gaurav asked as his interest intensified.

"Obviously, the hunt for the Ether started," I told them.

"How could the Ether be found? There is nothing visible in outer space." Mr. Singh asked

"When you travel in a train with its windows open, you feel a strong breeze even if there is no wind blowing. This shows that when an object moves through any medium, it creates a wind of that medium by imparting a part of its own kinetic energy to that medium," I started explaining.

"Correct." Mr. Singh understood and agreed with me.

"In the case of Earth moving through Ether, an Ether wind was supposed to be created at the surface of Earth. Speed of light in that wind was proposed to be reduced by an amount equal to the resistance provided by the Ether wind. And in reverse direction, this wind was believed to assist the propagation of light. Perpendicular to the direction of the Ether wind, no effect on the speed of light was expected. Then, for traveling a certain distance in the direction parallel to the direction of the Ether wind and returning back, light would take more time than traveling back and forth for the same distance in a perpendicular direction to the Ether wind."

"Why?" asked Mr. Singh.

"Take this example, where you are traveling by a boat on a river that is five km wide and has a current of five km/hr. The boat speed is, say, 20 km/hr. If you make a round trip from point A to point B, five km downstream and back, the net speed while going downstream will be 25 km/hr, and while returning, the net speed will be 15 km/hr." I made a diagram on the notepad provided to us as course material by the Institute.

River Analogy

Boat rowing across the river and boat traveling parallel to current

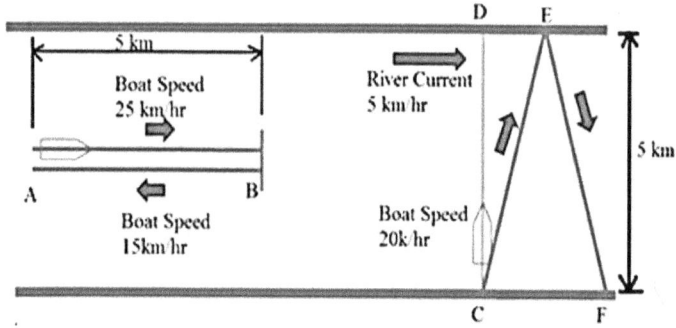

Figure 5
River Analogy to explain the difference in time taken in traveling along the current and perpendicular to it

After that, I asked Gaurav, "How much time will the boat take in traveling from point A to point B?"

He made a mental calculation and replied, "Five km distance at a speed of 25km/hr will be covered in twelve minutes."

"And from point B to point A?"

"Speed will be 15km/hr, so the time taken will be 20 minutes."

"Then, the total time for the trip will be 32 minutes. Right?" I asked.

"Yes," he agreed.

On the right hand side of the same diagram I depicted a boat going across the river and asked again, "In the second case, if you are going across the river at the same speed of 20 km/hr from point C towards point D, what will happen?"

"When going across the river, the current will be perpendicular to the direction of the boat, so that should not make any change in the speed of the boat, but the current will make the boat arrive at a point somewhere downstream."

"Right, you will reach point E instead of point D, but how much time will you take to reach point E?"

"To cover five km, it should take exactly fifteen minutes at a speed of 20 km/hr," he replied correctly.

"And from E to F?"

"Again fifteen minutes."

"Then the total time taken for the boat to come back to the same bank, although at a slightly downstream position F from the original position C, will be thirty minutes, Right?"

"Yes."

"Thus, there is a difference in time taken by the boat for a round trip parallel to the current, and perpendicular to the current."

"Okay."

"Earth moves through space at approximately 108,000 km/hr. It was thought that an Ether wind created by this movement should also cause a time difference in the traveling of light along the direction of the motion of Earth, and perpendicular to it."

"Good logic," commented Mr. Singh.

"Then Maxwell thought that if a time difference in the travel of light parallel to the Ether wind and perpendicular to it can be found, that will prove the existence of Ether."

"Then what happened?" Gaurav asked.

"An experiment was devised to measure this time difference. Two scientists Michelson and Morley made an apparatus of two long tubes, one extending the length of the room and the other vertical, making a right angle to each other. At the meeting point of the two tubes they placed a half-silvered mirror, so that when a light ray fell on the silvered part of the mirror, it got reflected, and if it fell on the other half, it could go straight." I made a diagram showing the arrangement.

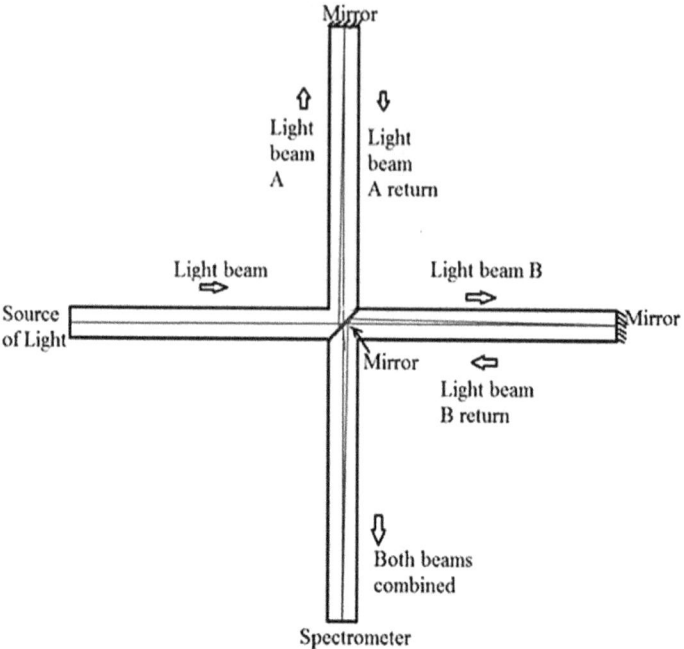

Figure 6
Experiment to measure the difference in time taken by light for traveling along the Ether current and perpendicular to it

"Okay," Gaurav muttered.

"Then, a source of light was placed at one end of the horizontal tube as shown. Light from that source got parted in two perpendicular directions depending on whether it fell on the mirror part or the transparent part of the half-silvered mirror, which was kept at a 45 degree angle to the direction of the light beam."

"So, one beam of light got divided into two beams; one horizontal and the other perpendicular," Gaurav repeated to prove he was following correctly.

"Yes, you are correct. The ends of both tubes were fitted with mirrors so both the beams reflected back," I told him further.

"Okay. Now the horizontal beam is traveling back toward the source, and the vertical beam is traveling downwards,"

Gaurav understood it correctly.

"The half-silvered mirror at the center of the meeting point of two tubes allowed the light coming from the top to traverse to the bottom of the tube without any hindrance, whereas the beam coming back horizontally got reflected by the mirror and its direction changed to vertically downwards."

"Thus both the beams combined again and their direction is vertically down now." Gaurav traced the route of both light beams with the movement of his finger on the diagram.

"The bottom of the vertical tube had a sensor which could sense whether one of the beams had reached it earlier than the other," I told them further.

"What did the sensor note?" asked Mr. Singh.

"There was no difference in the time taken by the two beams to reach the sensor."

"Did it mean that there was no Ether in the space?" Gaurav asked.

"Yes," I agreed, and continued, "Michelson and Morley had conducted this experiment very precisely underground to avoid any effect of other radiations. So, there was no chance of an error, and it was accepted by all scientists of that era that there was no Ether in space."

"If it had been proved very precisely that there is no Ether in space, then how can you say that their proof has some doubts?" Gaurav recollected the reason for discussing the experiment.

"Due to Lorentz's contraction," I supplied.

"What is that?"

"Lorentz was a famous scientist who had given the Lorentz transformations that were used by Einstein in his Theory of Relativity. He had hypothesized that when an object is moving at a speed comparable to the speed of light, its length reduces."

"Can you explain it a bit more? This seems an absurd statement to me. As per this statement, an airplane would shorten in length during flight." Gaurav was quick to make a comparison.

"That is right, an airplane will reduce in length during

flight, but this reduction will be so little that it will not be noticed by anyone. If the airplane flies at a speed comparable to the speed of light then it will certainly make a difference."

"Ha! ha! ha!" Gaurav cackled.

"What happened? Why are you so amused?" asked Mr. Singh.

Gaurav spoke up while still laughing, "Then God save the people sitting in economy class. Leg room for them will shorten further." This caused us also to guffaw.

"Don't worry; the passengers in economy class will also become thinner due to the contraction in the direction of the plane's velocity," I clarified after the laughter subsided.

"What is the effect of Lorentz's contraction on that experiment?" Mr. Singh asked on a serious note.

"One of the tubes lying along the motion of the Earth would shorten in length. Reduction in the speed of the light beam due to the presence of the Ether wind in that tube will get offset by the reduction in the length of the tube. So, there will not be any measurable disparity, even when the speed of light in both beams differed."

"Then, it is possible that Ether exists?" Gaurav asked excitedly.

"Yes, it is possible," I replied.

"Then why do the scientists use the Doppler Effect for explaining the red-shift of light instead of attributing it to the existence of Ether?" he wanted to know.

"Because Lorentz's proposition was not agreed upon. He had given his proposal after the experiment. It was believed that Lorentz had concocted the contraction merely to suit that experiment, so his suggestion could not be used to explain the result of Michelson's and Morley's experiment."

"If Lorentz's contraction was not used at that time, why you are using it now?" asked Mr. Singh.

"Because it's now been proven."

"Okay."

"There is one more fact I want to add. Michelson and Morley had never stated that Ether does not exist; they had only articulated that they could not find it. There is a difference in saying that you cannot find an object and that the object does not exist. The first case only shows your limitation of not finding the object. It should not be taken to be as equivalent to declaring that the object does not exist. The result from their experiment, though, suggested the first case, but was interpreted to be the second case. And the conviction kept on becoming stronger and stronger, as was the case with the relation of velocity of stars and their red-shift, which led to the theory of the Big Bang." I gave my view point.

"So, it is possible that Ether exists and causes the red-shift of light coming from the stars," deduced Mr. Singh.

"Yes. There is more proof for the existence of Ether that I'll give later, now we are reaching the guest house," I said as I saw the huge advertisement sign that had caught my attention that morning as we were leaving for the Institute. I was right; the van slowed down and took a left turn to enter the gate of our building.

At the guest house, Mr Singh and I watched television and talked about politics, while Gaurav surfed the internet. An influential leader of the ruling party had made a comment against a senior leader of his own party, and this news was being broadcast on almost all channels as it was the hot topic of the day. To gain political mileage, sometimes political leaders will go to any extent, even blaming their own party men. The ruling party had brought an ordinance to safeguard the political leaders from disqualification if they get a jail sentence from the court due to any crime committed by them. The government held the view that innocent politicians could also be framed this way by opposition parties, and would then get disqualified from entering the parliament.

The government wanted to protect them from disqualification. The matter of the ordinance was beyond the comprehension of the common people of the country, and there was widespread opposition to this act of the government. Sensing the mood of the populace, the President had sent this ordinance back to the government for reconsideration without signing it. Taking a cue from this, some of the leaders of the ruling party construed that the rug was being pulled out from under their feet, so in desperation, they blamed their own senior leaders and the cabinet for this ordinance, and called for its immediate withdrawal so that the opposition parties could not take advantage.

The news analysts were now giving their view on the latest development, and the clip-page of the young and influential leader asking for withdrawal of the ordinance was being shown repeatedly. We started getting bored watching the same news again and again, and decided to go for a walk.

While going down in the elevator, Mr Singh said, "I agree with your logic on the existence of Ether."

"Good to know that," I thanked him.

"What will be the effect of this Ether on red-shift?"

"We'll discuss that when Gaurav is also around," I suggested to Mr. Singh, as I wanted to avoid repeating myself.

"Okay," he agreed as we came out of the elevator. There was a security guard on duty at the gate of the building, who looked at us without any interest as we went out. There was a building under construction on the left side of us. After passing that, we came on to the main road and walked on the pavement, avoiding the potholes filled with rain water. There were small shops on the side where we bought some snacks and soft drinks. After roaming around for about half an hour, we returned. This time the guard stopped us at the gate.

"Which apartment?" he asked in a routine tone.

"1007."

He noted the number in his register and waved us in.

Apparently he had to make a log entry for everyone coming in, whereas he was not concerned for anyone going out of the building. That seemed strange to us, but he could have his own reasons, so we did not pay much attention and walked towards the elevator.

At the apartment, Gaurav was waiting for us. As soon as we entered the room, he complained, "Where had you gone?"

"We went for a walk," Mr. Singh replied.

"You should have called me also for a walk," he responded, still somewhat annoyed.

"We did not want to disturb you," I replied.

"No problem. I would have come."

"We shall take care next time," Mr. Singh assured him, handing over the pack of snacks to him. He calmed down.

"I read the course material," he informed us.

"Good. Then you can fill us in," I smirked.

He agreed and a discussion on the course material started, which got diverted to a narration of personal experiences while sailing. It continued during dinner, as well, and then went on until it was time to say goodnight.

The Big Bang and Lines of Space

Chapter 14
Perihelion of Mercury

The second day, the instructor dismissed the class one hour before the scheduled time. We called up Sunder, caretaker of the guest house, to send the pick-up van. After a few minutes, he arranged the van and informed us that the van was on its way. It stopped at the entrance of the Institute within ten minutes. The driver greeted us with a smile, as he had become familiar with us. Mr. Singh started talking to the driver as soon as he entered the van.

"How are you, man?" he asked.

"Fine, sir," the driver replied cheerfully and asked, "Your classes finished early today, sir?"

"Yes."

"That is good, sir. We'll not get stuck in traffic today; it is still too early for rush hour," the driver informed us.

The van was indeed doing better speed than the day before. Mr. Singh continued talking with the driver. They nattered on the subject of fuel prices that had been revised recently and had become the hot topic of discussion for everyone. The driver was blaming the government for the increase in fuel prices.

He commented with assurance, "This government will not win the next election; they have done nothing for the poor. Food prices have increased, gas cylinders have been limited, onions are going out of reach for even the middle class, and now the fuel prices increase." He shook his head. "How will people survive?"

"All this is happening because of corruption," Gaurav added.

"Yes, sir, all the politicians are getting rich and stashing their black money in foreign banks," the driver responded.

"If all the black money comes back, India will become rich

again," declared Mr. Singh.

"Then, maybe, I shall also own a nice car," the driver said and grinned sheepishly.

"Which car do you like?" asked Gaurav.

"I want to buy an Amaze," he replied.

"That is a good car; it has good mileage," added Mr. Singh. The first parameter of judging a car for most Indians is its mileage.

"At a speed range of 60 to 70 km/hr, it gives about 23-24 km per liter of diesel," the driver added.

"I like to drive at 90 to 100 km/hr on the highway," declared Gaurav.

"Sir, at higher speeds, there is more resistance from air; you don't gain the speed in comparison to the fuel burned," the driver gave his expert comment.

"Yes, that is right, but on the highway you can't drive less than that; other vehicles keep overtaking you. You have to maintain speed with them," commented Mr. Singh.

Suddenly, Gaurav was reminded of something and asked me," Sir, today, you have to give us proof of Ether."

"Yes, I remember. I will discuss it when we get to our apartment," I assured him.

We reached the guest house nearly two hours earlier than the day before, partly due to the cessation of the classes early, and partly due to less traffic on the way. Gaurav ordered Sunder to make tea, and without wasting any time, asked me, "Sir, how do you prove the existence of Ether, which can cause red-shift?"

"I will take you back to the time when the existence of Ether in space was disproved by Michelson and Morley."

"You have already told us about that experiment, and also the possibility that the medium of space can exist in spite of not finding any evidence of Ether," Gaurav recalled the result of the experiment.

"That experiment was mainly undertaken on the suggestion of Maxwell, who had thought that light waves traveled through a medium of space. Though he died before the results were in on this experiment, his idea lived on even after his death. The negative result of that experiment gave a setback to that idea. It became difficult to explain how light waves can travel if there is no medium of space."

"Then what happened?" asked Mr. Singh.

"Then it was thought that light, being an electromagnetic wave, could travel without a medium as the electric and magnetic waves will propel each other." I showed a diagram from my laptop and explained how a light wave traveling with speed 'c' will have electric and magnetic waves traveling with it and perpendicular to each other.

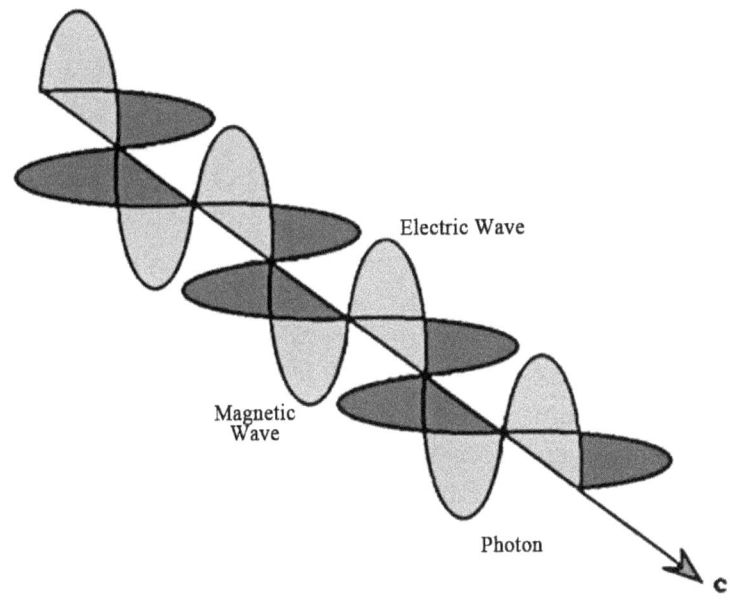

Figure 7
Electromagnetic wave with electric and magnetic fields perpendicular to each other

"Electric waves propel magnetic waves, and magnetic waves propel electric waves. Fantastic idea!" remarked Gaurav.

"Then light did not require a medium for traveling?" asked Mr. Singh in surprise.

"It did not. But for this idea to be true, it required the speed of light to remain constant." I added.

"There should not be any hitch with that. Speed of light is constant; it is always 300,000 km/second," offered Gaurav.

"It has to be constant for every observer," I cautioned them.

"It is constant for everybody." Mr. Singh reiterated Gaurav's statement.

"No, it is not. Let us consider a scenario where an astronaut is going away from Earth towards the Sun at a speed of 1km/sec. What will be the speed of light measured by him?"

"It will be 300,000 km/sec," replied Gaurav.

"But the astronaut is going towards the Sun at 1km/sec. Will he not see the light coming towards him at 300,001 km/sec?"

"Yes, he will," agreed Gaurav.

"Then how can you say that the speed of light is constant for all observers?" I asked.

"300,0001km/sec will be the result of two speeds, but the actual speed of light will remain 300,000 km/sec." Gaurav tried to stick to his point.

"No, the speed of light as measured by the observer has to be constant irrespective of his own speed. Since he will be observing the resultant speed, then he should observe it as 300,000 km/sec, not 300,001 km/sec," I clarified the condition required for electromagnetic waves to travel without a medium.

"How will he measure it to be 300,000 km/sec? It is not possible," opined Mr. Singh.

"It is possible only if we neglect the speed of the astronaut, as that will be very small compared to speed of light," suggested Gaurav.

"Let us assume that the speed of the astronaut is fast enough to be compared to light. Imagine it to be 100,000 km/sec, although that is highly unlikely," I said with a smirk.

"Then the speed of light measured by the astronaut will be 400,000 km/sec. It is not possible to still measure it as 300,000 km/sec," concluded Gaurav.

"You are right. It is not possible. A common man expects the speed of light observed by an astronaut to be the sum of his own speed and the speed of light when he is going towards the source of light. According to him, it is not possible for the speed of light to be constant for a stationary observer and a moving observer, as you have deduced from this instance. But scientists differed," I informed.

"Why?"

"In the quest to find a reason for light traveling without a medium of space, Einstein proposed that the speed of light can be constant for every observer if time is assumed to vary according to the speed of the observer."

"What do you mean by that?" asked Mr. Singh.

"When the astronaut covers a distance of 100,000 km and light travels 300,000 km in the same time, total distance covered will be 400,000 km. Now the resultant speed measured by the astronaut can be 300,000 km/sec only if the time taken to travel 400,000 km is slightly more than one second, because speed is equal to distance divided by time," I explained.

"Then there will be a problem, I think," Gaurav declared.

"What?" I asked.

"If more than one second is taken, then total distance traveled will increase further because the light will cover more than 300,000 km, and the astronaut will also cover more than 100,000 km." He put forward his point.

"No, as suggested by the scientists, time measured by the astronaut will still be one second, but that will be a longer second as compared to one second measured by the watches on Earth," I tried to explain.

"How can that be called one second if watches on earth record more than one second in the same duration?" objected Mr. Singh.

"That will be one second measured by the astronaut, which will differ from one second on the watches on Earth. Only to differentiate between the two types of seconds, let us denote the time in space as mecond (i.e. remove the 's' from second and add 'm' to indicate that it is measured by the moving observer instead of stationary observer)."

Both of my colleagues laughed at my newly devised unit of time, "mecond."

"Now the astronaut will measure the speed of light as 300,000 km/mecond, and he will cover a distance according to the time measured by him in meconds, not in seconds. He will have no means of differentiating between meconds and seconds, so he will consider them the same. According to the calculations, nine seconds measured on Earth will equal eight meconds measured at a speed of 100,000 km/sec. Thus the astronaut will cover a distance of 100,000 x 8 = 800,000 km instead of 100,000 x 9 = 900,000 km when he will travel at 100,000 km/sec for nine seconds," I explained the relation between two units of time.

"Will the distance covered by the astronaut be practically reduced to 800,000 km or it will be 900,000 km?" asked Mr. Singh.

"It will get reduced," I replied.

"You mean to say that the astronaut will actually cover a distance of only 800,000 km instead of 900,000 km in space in 9 seconds when he is cruising at 100,000 km per second?" Gaurav was still not satisfied.

"Yes, that is right," I verified.

"It seems a ludicrous idea," objected Gaurav.

"But this has been measured and proven beyond doubt. Scientists have actually measured the slowdown of time while traveling in a plane. It was done in 1971," I informed them.

"Strange! Time really slows down for moving bodies?" remarked Mr. Singh in bewilderment.

"Yes. It is true that this is a seemingly illogical idea for common people like us, but it has been proven and it helped in solving the most difficult problem of physics of that time. Therefore, all the scientists had to accept this idea, even though they opposed it initially," I told them.

"What was that problem?" asked Mr. Singh.

"Precession of the perihelion of Mercury."

"What is that?" both of them asked almost simultaneously and looked at me expectantly.

"All the planets move in elliptical orbits around the Sun and pass through the nearest point to the Sun once in every revolution; this point is called perihelion. Except for Mercury, every other planet has its perihelion fixed, whereas Mercury's orbit and, hence, its perihelion, shifts every time it completes one revolution around the Sun. This phenomenon is called precession of the perihelion of Mercury. This shift is very small – it is about 43 arc seconds every revolution (*an arc second is 3600th part of one degree angle*) and it takes 200,000 years for the perihelion of Mercury to complete one revolution around the Sun. In other words, we can say that in 200,000 years, Mercury makes one additional revolution around the Sun than it should, if the path of Mercury followed the pre-calculated path exactly as per Newton's Laws of motion," I explained with the help of Figure 8.

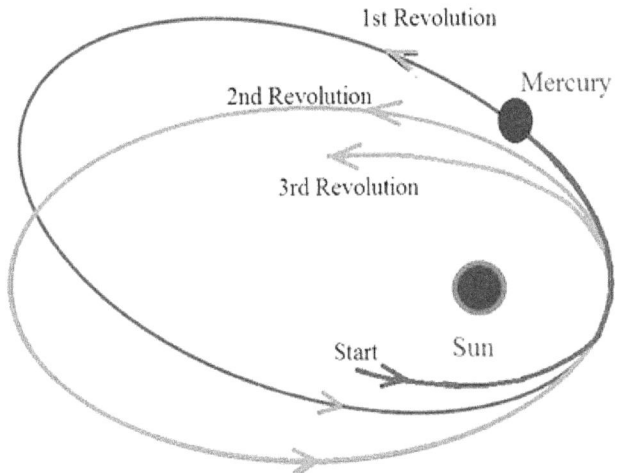

Precision of Perihelion of Mercury

Figure 8
Advancement of the orbit of Mercury showing precision of perihelion

"This is a very interesting fact. What is the reason for this shifting of Mercury's orbit?"

"When the planets are revolving around the Sun, we can calculate their exact position after any interval of time using Newton's Laws of motion; because we know all the forces acting on them, and we know that there is no obstruction on their way around the Sun. But, when Mercury completes its revolution around the Sun, we cannot calculate its exact position; there is a slight deviation in its path, which causes it to take a sharper bend when it passes near the Sun. Even after taking into account the effect of all the forces from the gravitation of other planets, scientists could not account for this deviation. When Einstein used the idea of variable time for moving bodies, and higher curvature of space near the Sun, to calculate the shifting orbit of Mercury in November 1915, he found that his calculations exactly matched the observation."

"Oh, that is why Einstein is so famous!" exclaimed Mr. Singh.

"That is only one of his contributions to science," I told him.

"How did Einstein use the idea of variable time for solving the problem of Mercury?" Gaurav asked.

"He reasoned that the time noted by someone on Mercury when it moved at very high speed near the Sun would be less than measured by another person on Earth, or you can say it will have longer seconds (or meconds), as I told you earlier. If this time difference and the curvature of space are taken into account while calculating the path of Mercury, it will give the exact path as observed. Thus, he showed that this time difference and curvature of space are the cause of advancement of its perihelion."

"I think I need some clarification here before we can proceed further," Mr. Singh restricted me from moving on.

"I'll give you one example. When you are riding a motorbike around a bend in the road; if the bend is sharp, you reduce your speed, don't you?" I asked.

"Yes."

"In other words, at slow speed, you can take a sharper bend."

"Yes."

"Similarly, you can imagine that Mercury takes a sharper bend due to its slowdown in speed at the turning, and starts its next revolution from a point slightly more advanced than it would normally do due to Newton's Laws of Motion. As the roads at the bend are slightly curved upwards at the outer edge, you can consider space to be curved at that point and thus assisting the faster turning of Mercury," I explained.

"Wait. I can imagine the curvature of the road and understand that the turning radius can be reduced for any vehicle at that point, but space is not a physical object; it cannot have curvature, it seems another illogical idea," objected Gaurav.

"Yes, this is the second irrational idea found in this theory besides the variation of time for moving objects," I agreed.

"They seem illogical to me also," declared Mr. Singh.

"Both of these ideas seem illogical, but their combination is logical. This combination solved the problem of the perihelion of Mercury," I commented.

"But, individually, the idea of the constant speed of light for all observers, and curvature of space, should also be logical," expressed Gaurav.

"If you want to explain the behavior of Mercury considering the absence of Ether, you will require these ideas; whereas in the case of the existence of Ether, you don't need any of these irrational ideas," I told them.

"How?" asked Gaurav.

"Let us take an example of a fast-moving train. As the train moves forward, it compresses the air in front of it, thus pressure of air rises and exerts force on the train so as to slow it down."

"Yes, that is right."

"This rise in pressure is directly proportional to the speed of the train?"

"Correct."

"Then the force acting to slow down the train is directly proportional to its own speed, Right?"

"Right," agreed Gaurav.

"Even the driver of our van knows that at higher speeds there is more resistance on his vehicle from air." Mr. Singh recalled the van driver's comment earlier in the day when we were returning from the Institute.

"Absolutely," I agreed with Mr. Singh's remark and then addressed Gaurav, "Now, imagine you can't see the air, or even know that it exists, then how would you account for the reduction in the speed of vehicles due to this resistance?"

"Must be a ghost," Mr. Singh suggested jokingly.

"Yeah, if nothing can be seen causing the reduction of speed, then it will be assumed to be the act of a ghost." I concurred with Mr. Singh with a smirk and added further, "Similarly, there is resistance from invisible Ether, which causes a slowdown of the moving objects. The reduction in the speed of the object due to this Ether is very feeble at low speeds, but as the speed of the object becomes comparable to the speed of light, the effect becomes more pronounced, and it will seem as if time has slowed down."

"Possibly," commented Mr. Singh.

"Since we cannot see this Ether, its resistance is not accounted for in Newton's Laws of Motion. And this slowdown caused by invisible Ether is similar to stating that time for the moving body has slowed down. Thus, we can calculate the distance covered by fast-moving objects by using Newton's Laws of Motion also, if we take into account the resistance offered by invisible Ether, or the ghost, as *Batti Sahib* says," I explained.

"This seems a rational justification for the slowdown of time for moving objects," commented Gaurav, and added after a slight pause, "It also provides good logic for existence of Ether."

"Thanks," I said.

"Now, we can understand the slowdown of time for moving objects, but are not yet clear about the curvature of space, which causes a sharp turning of Mercury as it passes near the Sun, like the curvature of a road," Mr. Singh remarked.

"I'll explain that later," I said and suggested that we should go for an evening walk.

Chapter 15
Bending of Light

There were no gardens near the building where we were staying, so we walked on the pavement along the road. On the opposite side of the road, we noticed a huge procession coming towards us. There was the noise of beating drums and blowing trumpets. From the sound, we could make out that a religious procession was underway. I asked my colleagues if they had any idea which god they were honoring. Mr. Singh, who had spent many years in Mumbai during his employment with the Indian Navy, guessed it to be an idol installing procession of the Hindu God Ganpati, as he recalled the approximate date of this event every year. As the procession drew near, it became clear that Mr. Singh was right.

Ganpati is also known as Shri Ganeshji, and is the most revered and loved deity of Hindus in this part of the country. He is such an important deity that he is worshiped by all Hindus wherever they reside. For starting any auspicious ceremony, he is the first to be invoked and asked for blessings. People believe that they must not start any job without paying respect to him. This belief has become so deep rooted in Hinduism that the name of Shri Ganesh has become a synonym of the word "start".

This God is the son of the most powerful God of Hindus, named Shiva. Shiva is the God who always wears a garland of snakes, and is dressed only in tiger skin. He has a trident in one hand and a *damroo* (a musical instrument like a small drum, which can be held in one hand and played) in the other. Nandi Bull is his vehicle, who, as some people believe, supports the Earth on one of his horns.

As a child, I would always become afraid every time I saw his photo on a calendar in our home. Although he had his beautiful wife and son standing next to him in that picture, he appeared scary due to the garland of snakes. The child in the photo had an elephant head. I remembered asking my grandmother why that child had that type of head. She told me a story which I never forgot. I don't know whether the story is true or not, but I believed it for most of my childhood until I started questioning everything scientifically.

God Ganesha was born as a perfectly normal child. His father Shiva had gone for meditation within the peaceful environment of the Himalayan mountains when Ganesha was only a child, so he did not remember how his father looked. When Shiva returned after many years from his religious excursions, Shiva did not recognize his son, nor did Ganesha recognize his father, which caused confusion when Shiva wanted to enter into his own abode. Young Ganesha did not allow his father to enter the house because his mother was taking a bath, and she had instructed him to guard the entrance.

There ensued a battle in which Shiva dislodged the head of his own son. When Gauri, the wife of Shiva, heard the noise of swords clanging, she came running out of the house, but by that time, it was too late; Ganesh was lying dead on the floor, without a head. His head had flown away with the immense force of the sword of Shiva. Gauri blamed Shiva for killing their only son and cried. She asked Shiva to make her son alive again. Not knowing what to do, Shiva invoked Brahma, the creator of the universe according to Hindu mythology.

Brahma appeared and heard Shiva's plight. It was a gaffe of immense proportions by Shiva to have dislodged the head of his own son, but Shiva had worshiped Brahma for so many years that Brahma was forced to give him a second chance. But the head of Ganesha could not be located. It would also have been so damaged wherever it landed that it could not be put back on the body. So, Brahma advised Shiva to bring the head of any child, and he would fit that on to Ganesha's body and bring him back to life.

Brahma put a condition on selection of the child, that Shiva should chop off the head of only that child whose mother he finds sleeping with her back towards the child. Shiva went searching the whole world but could not find any child whose mother slept with her back towards the child. In frustration, he started checking the animals also. He found one female elephant sleeping with her back towards her child. Shiva chopped off the head of the elephant child and brought it back to Brahma, who put the head on the lifeless body of Ganesha and gave him life. I believed that story when I was child, and I think millions of people still believe that story.

This story runs like a flashback in my head whenever I am reminded of Shri Ganesha, and my head bows automatically in reverence to him. I believe that even if that story is not true, there must have been something extraordinary about Shri Ganesha that caused so many people to worship him and love him.

The procession had gotten closer to us, people were dancing to the tune of the band and happily carrying the idols of Ganesha to their homes. They held small idols in their hands, whereas big idols were loaded onto trucks. Big idols were to be installed in public places in huge *pandals* (temporary structures) constructed specifically for the idols. Small idols were for installing inside homes. People would worship them for ten days. Wherever an idol would be installed, at least one person had to be in attendance of the idol, day and night.

After ten days, they would collect all the idols, take out a similar procession once again, and dispose of the idols in the sea. There, the idols would drown and the clay would melt in the water, while other ornamental objects would either decay or float on the sea, depending on their depleting nature.

As the procession passed by us, we watched it; young and old, men and women, all were dancing happily to the tunes of devotional songs. There seemed no worry whether they wore good clothes or not; everybody seemed to enjoy the coming of Ganpati.

We talked about the festival while we walked.

When we were back in the guest house, Gaurav reminded me, "Curvature of space…for bending the path of Mercury."

"There is no curvature of space," I said.

"What?" Gaurav asked in surprise.

"Then it is good…as it is, we can't visualize space to have curves," Mr. Singh remarked.

"Yes, if Ether can be proven to exist, then there is no requirement of considering the space curved," I clarified.

"You have already proven the existence of Ether by explaining the cause of the slowdown of time for moving bodies," Gaurav pointed out.

"That is only one idea. There may be some slip-up in that, so we need other proof for the existence of Ether, only then can we be sure about it," I told him.

"Where do we get the second proof?" asked Mr. Singh.

"Right here. If Ether can explain the bending of Mercury's path without the curvature of space, then that should be proof enough," I suggested.

"That is correct. But how will you prove that?" asked Gaurav.

"Using the bending of light," I said.

"But light travels in a straight line," Gaurav reminded me, as he had learned this in school.

"That is true, but it also gets bent by gravity."

"How can gravity bend the light? Light has no mass," objected Gaurav.

"Light coming from stars bends when it passes near the Sun." I put forth this fact.

"How can we be sure that light bends when it passes near the Sun?" Gaurav was still not ready to accept this as fact. In a way, his question was good for me; then I could clarify his doubts and prove to him the existence of Ether.

I informed him, "This bending was measured in 1919, when Einstein had suggested it."

"Really! Already done?" Gaurav asked in surprise.

"Yes."

"How?"

"Measurement of the bending of light was carried out because Einstein had predicted this phenomenon based on his

Theory of Relativity. The measurements of the degree of the bending of light matched with his prediction."

"Einstein predicted this bending of light before its observation?" Mr. Singh asked in surprise.

"Yes."

"That is strange," commented Gaurav.

"He became famous when this prediction made by him in 1915 was proven in 1919," I informed them.

"How did he make this prediction?" Gaurav was interested to know

"He used a thought experiment. Imagine that you are in an elevator enclosed on all sides, and the elevator is taken to outer space where there is no gravity; how much will you weigh?" I asked him.

"In space there will be no weight because weight is only felt due to gravity," he replied correctly.

"If the elevator is moving upwards with an acceleration of 9.81 m/sec^2, then you shall feel your weight the same as it is on Earth, right?"

"Yes," agreed Gaurav.

"Then, how will you differentiate whether your elevator is on the surface of Earth or moving in space?" I asked.

"We shall know when we are on Earth and when we are in elevator. The elevator in space will be moving very fast," suggested Gaurav.

"The elevator is enclosed from all sides, you will not be able to see outside, so how will you know whether the elevator is in space or on Earth? You can also imagine that the motion of elevator in space is smooth. There is no disturbance anywhere," I added.

"Then there will be the sound of the elevator movement and the air pressure difference," suggested Mr. Singh.

"Disregard the sound and air pressure difference also," I imposed another condition and added," In fact, there is no air in outer space, and sound cannot travel when there is no air."

"Then there is no way to distinguish between the two situations," resigned Gaurav.

"Einstein also said the same, and thus, he related gravity with accelerated motion," I advised.

"But gravity has the same unit as acceleration. What was so special in relating gravity and acceleration?" Gaurav asked as he did not understand the significance of this simple thought experiment of Einstein, so I had to explain it further.

"Imagine a source of light on one wall of this elevator when it is moving upwards at an acceleration of 9.81 m/sec^2."

"Okay."

"If a beam of light travels from this source directly toward the opposite wall, will the beam of light hit the exact opposite point on the other wall or it will hit slightly lower than that?" I asked.

"Since the light source is moving with the elevator, the light beam will be moving along with it and will hit at the exact opposite point," replied Gaurav with a rationale.

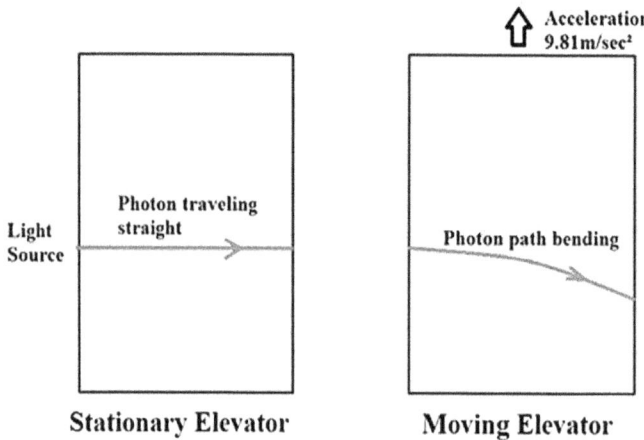

Figure 9
Bending of light in an accelerating elevator

"When the light leaves the surface of one wall, it will not remain under acceleration after getting detached from the source. It will move up with the speed at which it left the source, whereas the elevator speed will increase due to its acceleration. Hence, by the time light hits on the other wall, there will be some difference in the upward speed of light and that of the elevator, and thus the light photon (a unit of light) will hit at a slightly lower point on the opposite wall," I explained with a diagram.

"Got it," acknowledged Gaurav and Mr. Singh simultaneously.

"Now, using the theory of Einstein, where he articulated that acceleration and gravity are equivalent, what would happen to the light path when the elevator is at rest on the surface of Earth?" I asked them to make a comparison.

"In that case also, the light should fall while going from one wall to the other," replied Gaurav as he understood that the situation of an elevator on Earth was exactly same as the situation of an elevator in space under acceleration of 9.81 cm/sec^2.

"It suggests that light is being pulled down by gravity," I concluded.

"Oh yes!" Gaurav exclaimed as he realized the significance of this simple imaginary experiment.

"But there was one problem with this," I cautioned them.

"What problem?" asked Mr. Singh

"It needed proof, only then could this idea be considered by the scientific community. Any theory is considered valid only when it predicts something and then the observation proves that prediction," I advised.

"Why? If the theory is mathematically proven, then that should be sufficient proof, shouldn't it?" Gaurav objected.

"Let us consider an example. A software engineer claims that he has created a software program, where if he enters the number of customers coming to a shopping mall for ten days, then it will calculate the number of customers visiting the shopping mall on the eleventh day. This would be very attractive software for the shopping mall owner to plan his next day's sales. But before he buys the software, he asks the software engineer to prove that his program works.

The engineer very smartly takes the data of the last eleven days from the shopping mall, inputs the data for ten days and gets the value correctly on the eleventh day. But, the mall owner is smarter; he tells the engineer to predict the number of customers not for the days that have already passed, but for the days that are yet to come. That is the actual test of correctness of his program, where he does not have access to the data of the eleventh day. Similarly, a theory is correct only when it predicts something that is not yet observed, and when the observation is made, it matches the prediction. Einstein was looking for such proof. But his theory was difficult to prove in a laboratory."

"Then what did he do?"

"He thought if he found somewhere with enough strength of gravity to bend the light, such that the bending could be measured, only then could he get proof of his theory. But the gravitation of Earth is so little that any bending of light in this way cannot be measured. So, he looked towards the sky and found the Sun as a perfect candidate for proving his theory. He calculated the degree by which the gravitation of the Sun should bend the light coming from stars. He calculated it to be 1.74 arc seconds when the light passed nearest to the Sun, and proposed that if anyone measured the position of stars in line with the Sun, then compared it with the position of the same stars when the Sun was somewhere else, the position of the stars would be found shifted by 1.74 arc seconds."

"That was a very intelligent prediction based on theory and calculations only," commented Mr. Singh.

"Yes, it was. But it was very difficult to measure even this difference in the position of the stars in order to verify his theory."

"Why?"

"It is impossible to take a photograph of the position of a star when it is in line with the Sun. The star is not visible in the brightness of the Sun, although other positions of the stars could be photographed easily during the night."

"So how did they do it?"

"Astronomers had to wait for a total solar eclipse, when the sky near the Sun could be photographed and the position of the stars could be compared. In 1919, two independent teams went to take photographs of the total solar eclipse observed in Brazil and Africa. The observations made by these teams proved that Einstein was right."

"Unbelievable!" remarked Mr. Singh.

Gaurav exclaimed, "Really genius!"

"The bending of light around the Sun, found in 1919, was proof of the correctness of his Theory of Relativity. In his theory, he had said that time varies for moving bodies and space gets curved by gravity, which was not fully accepted by the scientific community until then. He was facing a lot of criticism, but after this finding, all the criticism died down and he became a hero in the scientific community. People readily accepted his Theory of Relativity, which he had used to predict this bending.

In his theory, besides the variation of time for moving bodies, he had articulated that time is a fourth dimension, and there is a space-time fabric in the whole of the universe. This space-time fabric gets warped due to gravity, and hence the light passing though warped space gets bent. The higher the gravity, the greater the bend. By the same logic, he could say that the space-time is warped near the Sun, so the path of Mercury gets bent and its perihelion advances by a fraction of a degree every time it passes near the Sun." I explained the significance of the finding.

"But, you had formerly advocated that there is no time variation, it is only a reduction in the speed of moving objects due to the existence of Ether," Gaurav pointed out.

"That is only if we assume the existence of Ether, but Einstein gave his theory without using Ether. So, he solved the biggest problem of his era by explaining the precession of the perihelion of Mercury, and non-existence of Ether by connecting them through his Theory of Relativity."

"It was really brilliant," Mr. Singh said appreciatively.

"Yes, it was. But, if we can prove the bending of light and bending of Mercury's path without the use of the fourth dimension of time, as we have already seen that variation of time for moving bodies is only resistance to speed by invisible Ether, then we can say that Ether exists." I gave away my way of thinking.

"Yes. That will also be brilliant if it can be done," Gaurav added with a smirk.

"We shall try to do that after dinner," I suggested as I noticed Sunder arranging dishes on the dinner table. I picked up the remote control for the television and switched it on. Disturbing news of a child slipping into a bore-well was being broadcast live.

In villages, to retrieve water from the Earth, villagers get submersible pumps fitted deep under the Earth. These pumps are fitted through a hole dug out by a drill machine. These holes are about one foot in diameter and can be up to 500 feet in depth. Usually, the mouth of these holes could be easily covered by a big stone or a sand bag. But, due to negligence of the people concerned, they can remain open and unattended. Small children playing nearby could fall into the bore.

One such incident had happened about two years back when a boy of about four years old had fallen sixty feet in one such well. To rescue that boy, the Indian army was called in. They dug a parallel well sixty feet deep, and then dug a tunnel towards the bore-well. After two days of struggle, they could pull the boy out safely.

Apparently, no lesson had been learned from that crisis. This time, a three year old girl had fallen in a well about forty feet deep. Efforts were being made with a similar technique to save her. The rescue operation had been going on since morning. We felt sympathy at the plight of the child and were worried about her safety. We forgot our topic of discussion and got involved in the live reports of the rescue efforts going on.

We stayed awake until midnight hoping to learn about the well-being of the girl. Finally the rescue team reached the girl and brought her out of the bore-well. She was immediately transported to the hospital, but bad news soon followed; the child had not survived.

There was an almost unanimous demand from the eye

witnesses of the event that administration should be stricter. They demanded the owner of the bore-well be held responsible for the death of the child as an example to other such bore-well owners to prevent such accidents in the future. That night I went to sleep with the thought that if there was a machine which could make negative gravity, then the life of that girl could have easily been saved.

The Big Bang and Lines of Space

Chapter 16
Refraction of Light

The next day, our classes at the Institute went on for one hour longer than normal because of the practical class. When we came out of the lecture hall, the van and driver were nowhere to be found. When we called the driver, he informed us that he had come to the Institute and waited for about half an hour, but had to leave for his next assignment at the airport. So, we had to hire an auto rickshaw to take us back to our place. It proved a tiring journey due to heavy evening traffic, numerous traffic signals, and the humid and hot climate of Mumbai.

We reached the guest house and went straight to our rooms, switched on air conditioning, and relaxed. Sunder was kind enough to leave us tea in our rooms. After having tea, though I tried to keep awake while surfing the net, I dozed off. I woke up when Sunder knocked on the door and told me that dinner was ready. I got up, freshened myself up, and went to the dining room. My colleagues were waiting for me at the dinner table.

"Looks like you had a good sleep," said Mr. Singh as I approached the dining table.

"Yes. I slept for about two hours," I responded while looking at the clock on the wall and making a mental calculation of the time.

"Sir, we are waiting for you to resume that story," Gaurav said.

"Which story?" I was still feeling a bit drowsy and did not understand what story he was referring to. I adjusted my chair so I could watch the news while I ate.

"About the bending of light by the Sun without the use of the fourth dimension of time," Mr. Singh reminded me.

"But that is not a story," I said while serving myself

vegetables from the bowl.

"That is like a story for us; it is interesting," Gaurav said.

"Okay," I said, but turned my attention to the television. There was no sensational news that day. Still, I kept watching while I quietly ate. My colleagues resumed their discussion on the new features of mobiles and the latest applications available after they sensed that I was not yet in the mood to talk.

After we finished dinner, we settled down on the sofa chairs. They had become quite cozy over the three days we had lived in the apartment. Now that Sunder was aware of what was expected of him, he was ready with tea as soon as we had finished dinner.

I felt better after having a few sips of tea, and suggested Gaurav reduce the volume of the television so that we wouldn't get disturbed during the story that I was going to tell them. He immediately complied.

"In fact, the story of finding the bending of light by the Sun had finished yesterday. Today, we just have to see whether this is possible or not if there is no fourth dimension of time, i.e. time does not vary for moving bodies," I commenced.

"You mean if the time variation theorized by Einstein is nothing but the reduction in the speed of moving bodies due to invisible Ether," Gaurav clarified my statement further.

"Yes, you are right," I agreed with him and resumed, "To prove the bending of light without the fourth dimension of time, we shall again assume Ether to be similar to air as we had done in the case of finding time variation for moving bodies. The only difference is that air is visible and Ether is not visible."

"Okay," Mr. Singh agreed.

"Air is dense at the surface of Earth, and as we go upwards from Earth, the density of air keeps on decreasing. Right?"

"Right," responded Gaurav.

"Why?" I asked.

"Simple; it is due to gravity. Heavier elements of the air tend to remain near the Earth, and the lighter ones are found in the upper atmosphere," Gaurav replied correctly.

"So, the density of air is dependent on gravity. Similarly, density of Ether should also be dependent on gravity. Ether should be denser where there is high gravity," I stated the comparison.

"That is reasonable to deduce," Mr. Singh agreed and indicated for me to go on by nodding his head in agreement.

"Thus, Ether, if it exists, will be denser near the Sun than away from it."

"Correct," Gaurav agreed with my logic.

"Do you remember refraction of light?" I asked him.

"Yes. When a light beam falls on an inclined transparent surface such as glass or water, it gets bent," he replied.

"Good. Do you remember the reason for that?" I asked further.

"Something to do with the density of mediums," he recalled, but could not give the exact answer.

"When light goes from a rarer medium to a denser medium, it bends towards the normal to the boundary of the mediums and vice-versus," I reminded him.

"Yes."

"Now, if there is denser Ether near the Sun, then the light of a star coming toward the Sun will bend towards the Sun, and when it goes away from the Sun, it will bend outwards."

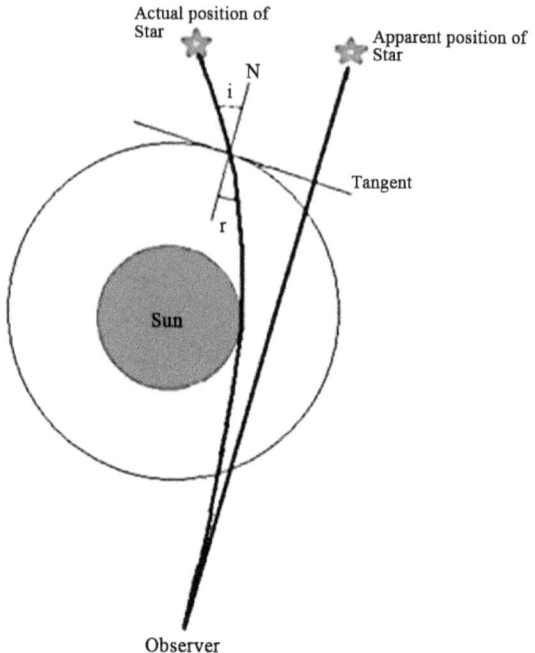

Figure 10
Light coming from a star, bent according to its property of refraction when passing near the Sun

I related the bending of light with the existence of Ether and showed with Figure 10 the bending of light through refraction due to different densities of Ether near the Sun. The circle represents the boundary between dense and rare medium. Tangent at the point of entering the rarer medium represents the inclination of the boundary to the incoming light ray. Perpendicular to the tangent is the normal 'N'. Incidence angle and refractive angle are represented by 'i' and 'r' respectively.

"Will it bend in the same way as shown by Einstein?" Mr. Singh asked for clarification.

"Yes, it will."

"You don't require space-time curvature for explaining this bending of light?" Gaurav asked, as he seemed bowled over by the simple logic of refraction of light by invisible Ether.

"No."

"Wonderful!" exclaimed Mr. Singh.

"It means that the existence of Ether is proven?" asked Gaurav.

"It does not prove the existence of Ether; it only suggests that bending of light by the Sun can also be explained with the existence of Ether," I gave my view point conservatively.

"No. I think Ether must exist," Gaurav said emphatically and added his logic, "The idea of Ether makes the variation of time and curvature of space redundant. It must be true."

"Yes, and it will also provide the medium for light waves to travel," Mr. Singh pointed out its third benefit.

"I agree with you that everything falls in place with existence of Ether. Light waves get the medium for traveling as Maxwell had suggested, Time dilation does not seem irrational, as this is shown to be caused due to resistance to motion of fast moving objects by invisible Ether, and thirdly, the bending of light is illustrated to be due to its own nature of refraction, not by gravity or curvature of space; I agree to all that, but we still need concrete proof of its existence," I advised, exercising caution.

"As far as I have understood it, Ether exists," declared Gaurav.

"Me too," added Mr. Singh.

"Thanks for your confidence in the idea of Ether, but before we become absolutely certain about it, we need to probe it further, we need to know its properties, and we have to collect indubitable proof, because non-existence of Ether was proved 137 years earlier. Since then, science has come a long way; we can't simply announce that Ether exists and everything should be changed accordingly," I advised.

"You are right. Then what is your plan?" asked Mr. Singh.

"I will study it further."

"How can you study something that you can't see?" asked Gaurav.

"Einstein also did not see the light bending around the Sun?" I pointed out to him with a smirk and got up. No answer was needed from him, so I said goodnight and strolled towards my bedroom.

Chapter 1

PART- III
LINES OF SPACE

Chapter 17
Lines of Space

The next day, we were scheduled to meet company officials after completion of our classes. The meeting, held at a coffee house, went well, and it was late in the evening before we got back to our guest house. Sunder was not at the guest house when we arrived. We called him up and found that he had gone to meet his friend in another apartment in the same building. We waited in the alleyway until he came. He was apologetic for making us wait outside the apartment and served us tea in our rooms right away.

To avoid falling asleep just yet, I picked up the tea cup and sauntered into the common room and switched on the television. The prime minister had just returned from abroad, and there was speculation of tendering his resignation in view of his cabinet's decision being thrashed by a younger member of his own party three days prior. Most of the news and analysis pertained to the hypothetical resignation of the prime minister. I wondered at the way unfounded speculations were made by overly ambitious journalists, and how debates were organized on news channels, while experts were invited to offer their opinions.

All this was being done to make big news for gaining popularity. Such news would then continuously broadcast on various news channels. They give so much publicity to the speculated news that an authority usually has to intervene to furnish a clarification. That clarification also becomes more news. So, they keep the people glued to television sets with conjured up news. This trick works for many news channels. I was certain that news of the resignation of the prime minister would prove to be such a trick.

Slowly, my mind drifted from news to science. Science also has its share of similar tricks. The existence of dark matter, dark energy, the theory of the Big Bang, and Inflation theory, it all seemed to fall into the same category of such tricks. They do not exist in reality, but have been debated so much that they give the impression of a reality. Unfortunately, in science, it is very difficult to either deny or prove these speculations.

I was lost in my thoughts when I was suddenly shaken by the touch of Gaurav's hand. He informed me that Mr. Singh was talking to me. I did not realize that both of them had come in and sat down. I did not know what Mr. Singh had asked me, so I was somewhat embarrassed.

Mr. Singh realized this, so he quickly explained, "Gaurav and I were discussing that if Ether is proven to be present, then even the Big Bang theory will fail. And if the Big Bang did not happen, then there must be an alternate theory for the creation of the universe." It was a question to me in the form of a statement.

"Brahma must have created everything," I said casually as I recalled the story that had flashed through my mind the day before.

"No, I didn't mean that, I meant scientific theory," he smirked as he understood my comment. In India, just about everyone knows that Brahma is the creator of the universe, according to Hindu mythology, and, hence, can relate immediately to my comment.

"One of the scientific theories is 'string theory'; you must have heard of that," I said.

"I have heard of string theory," replied Gaurav.

"Good, then you can enlighten us on that," I said.

"I only said that I have heard of it; I didn't say that I know it very well!" He backed out. Mr. Singh and I cackled at the way Gaurav reneged, and then Mr. Singh suggested, "Sir, only you can tell us about that."

"In string theory, it is assumed that everything is made of very small strings. Even electrons, protons and other sub-atomic particles are made of strings. These strings keep on vibrating. Their vibrational pattern decides the nature of a particle made by them," I told them.

"Why do they keep vibrating?" asked Gaurav.

"That is an assumption. Moreover, these strings require eleven dimensions to explain this theory."

"Even the fourth dimension is difficult for us. How can we visualize eleven dimensions?" Mr. Singh expressed his inability to go further on this theory.

"Sir, I think it would be better if we moved on to something other than string theory," suggested Gaurav.

"I can explain to you how to visualize eleven dimensions. They are curled into other dimensions," I offered.

"Curled dimensions? No, that must be very difficult."

"In that case, we have another theory, as I've already told Gaurav, empty space can convert into particles and anti-particles. It is like zero converting into +1 and -1. All the particles of +1 type form our universe, whereas all the other particles create some other universe, with everything in that universe made of antimatter," I said.

"Where is the other universe?" asked Mr. Singh.

"The physicists proposing this theory do not have an idea of the whereabouts of that universe just yet."

"It probably exists only in their imaginations. This idea does not seem realistic," commented Gaurav.

"Otherwise, leave off on the creation of the universe, tell us more about Ether," suggested Mr. Singh.

"First I will tell you how to visualize Ether. It will be in the form of lines, so I will call it Lines of Space."

"That is the name of your book," Mr. Singh recalled.

"Yes. I am going to use some figures from that book," I told them and brought out a printed copy from my bag, which I always carried with me.

"That is a nice cover," Mr. Singh commented as he saw the

book.

"You did not show it that day when you told me about this," complained Gaurav.

"It just didn't occur to me, and you never asked to see it," I pointed out.

"We didn't know that you were carrying the book," explained Mr. Singh.

"No problem. You can read it later. Now we shall resume our visualization of Lines of Space."

"Okay."

"Imagine a cubicle box of one meter sides, having graph paper sheets of 1 cm square divisions pasted on all six sides of the box. Each edge of the box will have 100 points and each side will have 100 x 100 points. If you connect all the points of the graph paper pasted on one side, with the opposite side's exact similar points through imaginary straight lines, you will get 10,000 lines going from one side to the other side. Since there are three pairs of opposite sides in the box, a total of 30,000 lines will be present inside the box. It will be like a three dimensional grid kind of arrangement." I made a rough sketch as shown in Figure 11 for visualizing the Ether, or Lines of Space, as I called them.

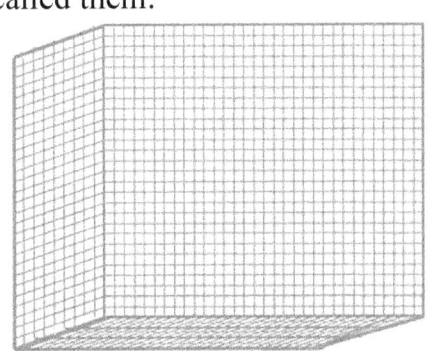

Figure 11
Uniform distribution of Lines of Space

"What is the thickness of these lines?" asked Gaurav.

"There is no thickness; they are imaginary lines and they will have much less gaps than appear in this figure, even less than a billionth of a millimeter."

"Hmm." Mr. Singh murmured, and indicated I should go on.

"Now, consider a sphere of radius R with this kind of medium of space as shown in Figure 12." I showed them a figure from the book and clarified to them that I had not shown the grid of lines of space in that figure because they were so close that the space would appear smooth."

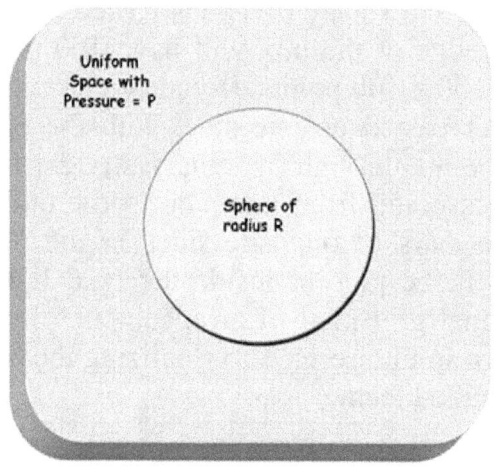

Figure-12
A selected sphere of radius R in uniformly distributed Lines of Space

"What is this uniform pressure P in the figure?" asked Gaurav.

"Until the time there is no disturbance in the Lines of Space, they will be exerting equal pressure in all directions similar to the pressure exerted by still air. Because we are considering Ether similar to air, so it has to behave that way," I explained.

"Okay." He understood.

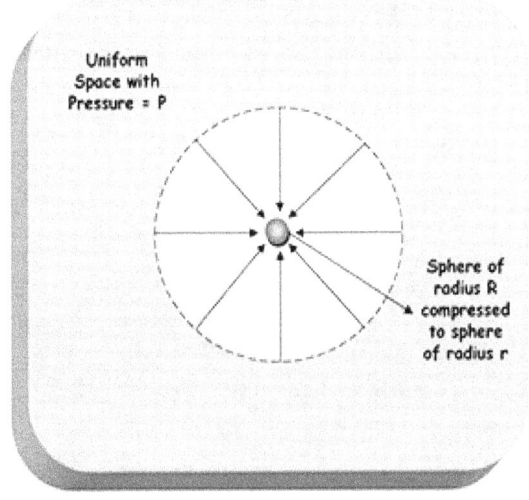

Figure-13
Contraction of Lines of Space in sphere of radius R into a smaller sphere of radius r

"Now, let us suppose that the Lines of Space inside the selected sphere contract to a sphere with a very small radius as shown in Figure 13. A vacuum will get created in the space between the original sphere and the contracted sphere. This vacuum will cause the Lines of Space lying outside our selected sphere to get pulled inwards."

"Why will the sphere get contracted?" asked Gaurav.

"This is an assumption, but it may happen due to a change in temperature."

"Will there be a sudden temperature change for the sphere to contract?" Gaurav was yet not convinced about the contraction of the sphere of the Lines of Space.

"There may be a slow change, or fast; it doesn't matter. I am only showing initial and final conditions." I made myself clear.

"Okay." He got it.

"This will cause the Lines of Space in the surrounding area to get stretched as shown in Figure 14. Darker shades show more stretching and lighter shades show less stretching."

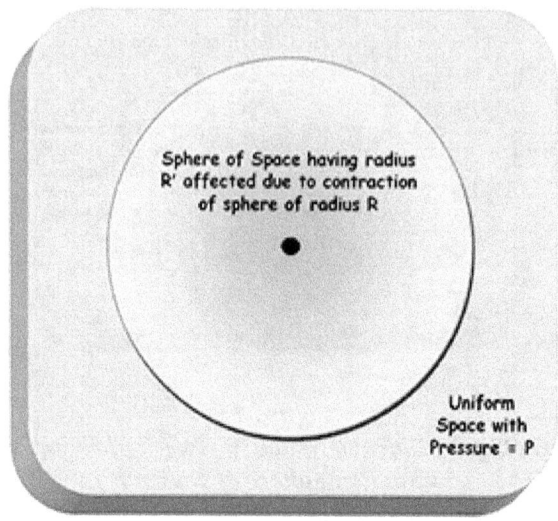

Figure 14
Stretched Lines of Space in a bigger sphere of radius R'

"But, this figure has a bigger sphere than the original sphere. Why?" asked Gaurav.

"Because the effect of this disturbance will extend further into outer space, beyond the original sphere. I have depicted all the space that will have stretched Lines of Space. Beyond this sphere, there will be unaffected space as the original."

"Okay."

"This spherical space with stretched lines will rearrange itself according to the stress in the lines. They will not remain straight lines going from one end to the other end as was the original case. Now the lines will become circular, having their center coinciding with the center of the compressed sphere, and there will be radial lines emanating from the outermost surface of the stretched sphere towards the center of the compressed sphere. This way, the space will become non-uniform, having different density and stress in the Lines of Space at varied locations as shown in Figure 15," I explained further.

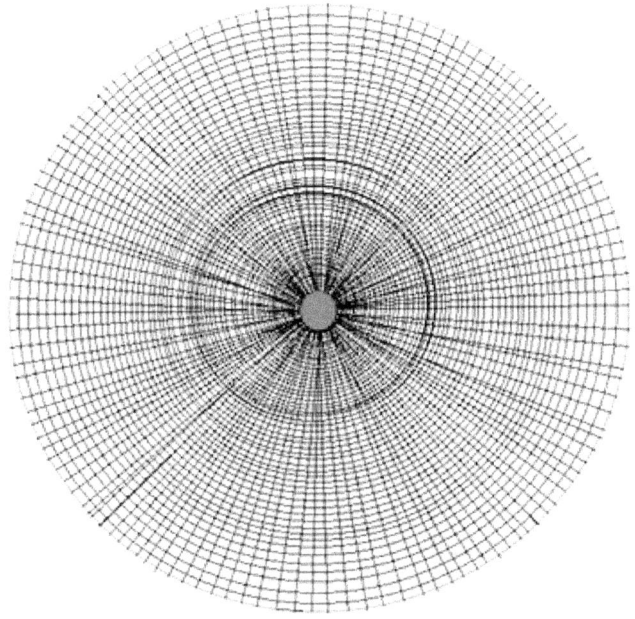

Figure 15
Re-arrangement of Lines of Space after contraction

"So, the initial uniform space will become like this after the contraction of a spherical part of the Lines of Space?" Mr. Singh asked.

"Yes. And now, I shall tell you about the formula to find the number of Lines of Space, and the stress on them at any distance from the center of the compressed sphere that is represented by a small circle at the center of Figure 15."

"I hope that is not too difficult," Mr. Singh said with a grin.

"I have derived this formula in my book; here I shall just mention the result. You can get the details from the book *'Lines of Space.'*

"Okay," agreed Gaurav and indicated for me to go on.

"Take a look at Figure 16. If the radius of total stretched space is assumed R', then the number of lines per unit area on the surface of the sphere of any intermediate sphere of radius

R" will be m", which can be found using the formula $(1/3) n (z^3 - 1) R^3 /\{(R")^3 (\log R' - \log r)\}$, where n, z, R , R", and r are as follows:

n = Number of Lines of Space per unit volume in uniform space

z = Ratio of the radius of the stretched sphere and original sphere

R = Radius of the original sphere of uniform space

R' = Radius of the sphere effected due to contraction of the sphere of radius R to the sphere of radius r

r = Radius of the contracted sphere

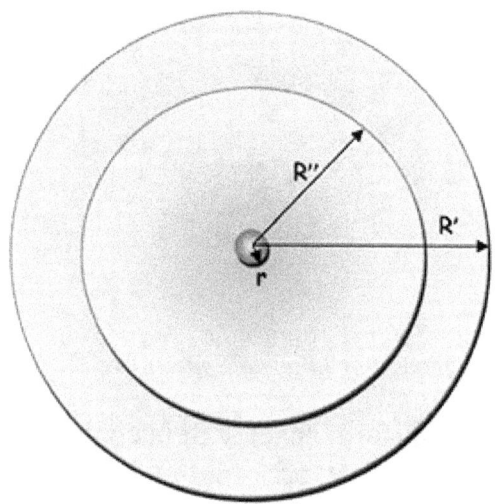

Figure 16
Any intermediate sphere of radius R" between spheres of radius R' and r

"It appears to be a complicated formula," remarked Gaurav.

"No, not too difficult. After finding the value of (m"), we can also calculate the stress (t") in each line at the surface of the sphere of radius (R") by using the relation, $m" = k/(t")^3$ where k is a constant. The product of these two quantities will give us the total stress on the surface of the sphere of radius (R")," I added further.

"What is the use of calculating this stress?" asked Mr.

Singh.

"This stress is inversely proportional to the square of the distance from the center of the contracted sphere."

"What does that show?" Mr. Singh asked again.

"It can be related to gravity," I told him.

"How?" asked Gaurav.

"Because gravity is also inversely proportional to the square of the distance from the center of a body, such as the Earth or the Sun."

"Just one similarity?"

"Yes. But this resemblance is very valuable."

"Why?"

"Imagine if the Earth only consisted of the nuclei of all the atoms used in its formation.

"No electrons?"

"No. Then it would shrink to a sphere of only about a 640 meter radius."

"Why?" asked Mr. Singh.

"Because the size of the nucleus of an atom is about 1/10,000th of the size of an atom."

"That would be very small. Then the circumference of Earth would only be about four kilometers." Gaurav calculated quickly.

"Then someone could even walk around it in one hour," Mr. Singh added with a grin.

"No, that would not be possible, because a man would weigh one hundred million times his present weight. I don't think he would be able to walk," I said jokingly, and both my colleague laughed heartily.

"Nobody would survive in that situation," Mr. Singh commented while still doubling up.

I waited for the amusement to abate and then resumed, "If that small size of Earth is assumed to have contracted from space, which is one billion times the volume of this compacted Earth, and then the stresses are calculated in Lines of Space at the surface of Earth, as per the formula derived, then we find that the result behaves exactly as the gravity of Earth. After equating the stress in Lines of Space at a distance of 6400 km from the center of the compacted Earth with the gravitational attraction at the surface of the normal size of Earth, i.e. 9.81 m/sec², we find that the value of gravity at any distance from the Earth is exactly the same as the stress in Lines of Space, calculated by using our formula." I explained the relation between gravity and Lines of Space.

"That is interesting," commented Gaurav.

"In short, we can say that if Lines of Space condense to form matter, then a stress in the remaining Lines of Space will be produced around the condensed matter, which will behave exactly like gravity." I repeated myself so that I could be better understood.

"You mean to say that all the matter on Earth could have been created from space itself?" asked Gaurav, as he understood my concept.

"Yes."

"It is amazing. I thought Ether, or this medium which you have named as Lines of Space, was assumed only for facilitating the travel of light waves," Mr. Singh expressed.

"It is much more than that. All the protons, electrons, etc. are created from these Lines of Space. This concept even gives the relation between gravitation and other fundamental forces. In this concept, we are not required to imagine the universe to have been created from a small point through a gigantic explosion or Big Bang," I explained.

That day, we got so busy in our discussion that we lost track of time until Sunder called us for dinner. During dinner, Mr. Singh declared, "I have a question."

"What is that?" I asked.

"If all the universe is made from the smooth transition of Lines of Space into matter as you advocate, and there was no Big Bang, then why did other scientists not contradict the Big Bang theory? How can all the scientists of the world be on the side of the Big Bang theory, and you alone, not being a scientist even, try to refute such a popular theory?" Mr. Singh expressed his feelings very honestly.

"Yes, that is the reason why I had been hesitant in agreeing with you until you gave logical reasons against the Big Bang theory," Gaurav agreed with Mr. Singh's query.

"I am not the only one against the Big Bang theory. This theory has faced opposition from the time it was proposed about eighty years back. First of all, it gave the origin of the universe to only be about 1.5 billion years, whereas Earth was known to be at least 3 billion years old. This was a known 'age problem'. When the timing of the Big Bang was pushed back to about 13 billion years with further research, there was a problem of its proof, so this theory remained stagnant until the finding of cosmic microwave background radiation (CMBR). But that radiation caused other problems, as I've already told you. This theory has faced opposition from many physicists throughout its journey. There are many techniques proposed by various other physicists too for opposing this theory; mine is only one of them," I replied.

"So, the Big Bang theory is not final; there is a possibility of change here," construed Mr. Singh, and I agreed with him.

Chapter 18
Sub Atomic Particles

After finishing dinner, we sat down once again in our respective sofa chairs. My concept on the Lines of Space was weighing so much on the minds of both my colleagues that they continued the discussion on that subject even after dinner. Mr. Singh had something on his mind he wanted to ask, but before he did, he summed up the discussion so far in his own style. "According to the concept of Lines of Space, the Big Bang did not happen, and the universe created itself smoothly from space by contracting the medium present in space, i.e. the Lines of Space."

"Right." I indicated for him to go on.

"Then all the sub-atomic particles, protons, neutrons, and electrons, also must have formed from space itself."

"Yes."

"How?" He put up his question finally.

"Contraction of the Lines of Space will give us protons, positrons and neutrons, while expansion will give electrons and muons etc."

"What about their charge?" asked Gaurav.

"Contraction will correspond to a positive charge and expansion to a negative charge," I stated.

"Then what about neutrons?" asked Mr. Singh.

"Neutrons will result from a combination of positive and negative particles." I furnished a simple solution.

But Gaurav was not satisfied, and declared his doubt. "In that case, both of the particles should annihilate each other."

"Yes, he has a point," Mr. Singh agreed with him.

"In a hydrogen atom, there is one proton and one electron, why don't they cancel out each other?" I asked Gaurav.

"They are two different particles having different mass," he replied.

"It's for that same reason they can co-exist to form a neutron also."

"Then they will form a hydrogen atom; there seems no other way that they can stay together," objected Gaurav.

"They can. But you shall have to learn about the behavior of Lines of Space in my book," I informed Gaurav.

"I shall read that, but can you explain it in brief?" he asked.

"In the Lines of Space concept, a proton is formed by contraction of space, whereas an electron is formed by the expansion of space, so the proton will be much smaller in size than the electron. It will then depend on the position of a combination of protons and electrons whether they will form a neutron or an atom. If an electron is revolving outside a proton, it will form a hydrogen atom, and if a proton goes inside the electron, it will form a neutron." I gave a succinct synopsis of the concept.

"But a proton cannot go inside an electron?" objected Gaurav.

"According to the concept of Lines of Space, it can. I told you that an electron will be much bigger than a proton, so the proton can go inside that."

"But an electron is the smallest sub-particle; how can you call it bigger than a proton?" asked Mr. Singh.

"According to the quantum theory of physics, you can never locate an electron exactly. Its position is always expressed in probability; that indicates that an electron is not a point particle, it is a cloud. The Lines of Space concept also proposes an electron to be a cloud type particle much bigger than a proton." I gave the comparison of the appearance of an electron with quantum theory.

"Maybe," resigned Mr. Singh as I delved into quantum theory.

"According to you, three types of particles can get created from the Lines of Space," Gaurav tried to summarize.

"No, only two types of particles will get created from space, and their various combinations will result in other particles and

The Big Bang and Lines of Space

atoms."I corrected him.

"This is not the way we learned about protons and electrons; this is totally different," expressed Gaurav.

"Yes. It is different. If you consider the existence of a medium of space, you will have to change the trend of physics from that time onwards, when the medium of space was discarded. The Theory of Relativity will change, quantum physics will change, and there will be no Big Bang.

"How will quantum physics change?" asked Gaurav.

"I just told you that the structure of sub-atomic particles will have to be reconsidered. That comes under quantum physics," I replied.

"That will be very difficult; these theories have already been proven, haven't they?" remarked Mr. Singh.

"They are not firmly proven, they also face plenty of problems, similar to the problems of the Big Bang theory. Therefore, one should not consider an idea impossible just because it contradicts earlier prevalent ideas. There is a possibility of change even after hundreds of years. Galileo changed Ptolemy's idea of the rotation of the Sun around the Earth after fourteen hundred years, Thomas Young changed Newton's idea of particle nature of light after one hundred years, and Einstein changed Newton's idea of gravity after more than two hundred years. If the Lines of Space concept provides better understanding of the universe, then that should be explored further," I advised philosophically.

"But the people who caused such drastic changes in science were great scientists, whereas you are not a scientist," Gaurav declared his doubts.

"We should judge the idea, not the person who is proposing it. I don't matter; it is the idea that matters," I replied.

"Lines of Space already seem better in explaining some of the concepts, so I think it should be given a chance," Mr. Singh agreed with me.

"I also agree with that, but the change that the Lines of Space concept is proposing is too big," declared Gaurav.

"If the change is making something better, then it doesn't

matter whether it is big or small, it should be explored," I suggested and finished the discussion for the day.

Chapter 19
A Comparison

It was the last day of our training. We had gotten ready to go to the Institute, and were waiting for the van. Usually, the van arrived in the portico of the building before we came down from our apartment every morning, but that day, it was late. After waiting for about five minutes, we called up the driver and found that the van had broken down while coming to the guest house, so he had arranged for another vehicle, which would take about fifteen minutes to reach us. We could hire a taxi and depart, but we decided to wait.

After a few minutes, a car pulled up near us. As soon as we entered the car, Mr. Singh asked the driver what had happened to the van. He started narrating about the breakdown of his fellow driver's van, and engaged my colleagues in the dialogue. I was not interested in their talk, so I started ruminating about my favorite subject.

Although the views of my colleagues regarding the Big Bang had undergone a complete reversal due to our discussion in the last few days, that was not sufficient for me. I had to prove my concept of Lines of Space beyond doubt. I had been working on doing just that for quite some time, but it was not easy. I, being a common man without the benefit of the intelligence of Einstein, had less of a chance of achieving that. But his one quote always kept lingering in mind. He had said, *"I don't have extraordinary intelligence, I only stay longer with the problem"*. Although, this quote was a sign of Einstein's humility, it kept me inspired to keep trying to find a solution. Whenever I had free time, I used to keep ruminating about that and tried to find faults in my own concept to make it better. I started comparing Einstein's Theory of Relativity and my concept of Lines of Space, while Gaurav and Mr. Singh were busy in conversation with the new driver.

Einstein had assumed that the speed of light was constant for all observers, irrespective of their movement. According to the Lines of Space concept, the speed of light would also be measured constant by all observers, irrespective of their movement. The only difference lay in the explanation. Einstein theorized the presence of an internal clock for each observer which measures time according to the observer's speed. This internal clock slows down as the observer moves and, hence, the net speed of the light measured by the observer remains constant. Whereas, the Lines of Space concept advocates a drop in the speed of light equal to the speed of observer due to the resistance from the invisible Ether, and, hence, the resultant speed of light measured by the observer is again retained invariant.

Einstein had made a second assumption that there exists a space-time fabric that curves at high gravity. The higher the gravity, the higher the curvature of the space-time, and vice-versus. Curvature of space-time was shown by him through the bending of the light of stars by the gravity of the Sun. On the other hand, the Lines of Space concept does not require linking space and time. It articulates that the medium in space is denser near the region of high gravity. Due to the high density of invisible Ether near Sun, the speed of light drops and it gets bent due to its own property of refraction, than due to the curvature of space. In this concept, there is no need to assume the curvature of space and fourth dimension of time, thus making it easy for the common man to relate with the reasoning of bending of light by Sun.

Thus, according to my estimation, Lines of Space seemed to be a better alternative than other theories on the subject due to its simplicity. But I also knew that I could be admonished for even thinking that. It looked as if I was working against the theories proposed by the great scientists, but on close scrutiny, invisible Ether or Lines of Space could be considered equivalent to Einstein's space-time fabric.

I wondered if there existed no medium in space as was

advocated by Michelson and Morley's experiment, then what was the space-time fabric of Einstein, was it not a medium of space? And what is the Higgs field as advocated by Peter Higgs in quantum physics? Is it also not some kind of medium present throughout space? As two famous scientists – Sir Albert Einstein and Peter Higgs, had both advocated that something is present in space, then why not consider it Ether, or Lines of Space, and set things right from the time when Ether was discarded? Yes, there has to be a medium of space present in the entire universe.

If the presence of a medium of space had not been discarded, the Big Bang theory would have been abandoned in the 1930s when Lemaitre had suggested it for the first time. It got support from Hubble's Law of receding galaxies based only on the Doppler Effect. Thereafter, Einstein's Theory of Relativity helped the Big Bang theory. He had assumed gravity to be the only force present in the entire universe. He reasoned that the entire universe should collapse due to such high gravitational force. To avoid such collapse, there must be some other unknown force that is preventing it from collapsing due to widespread gravity.

In his Theory of Relativity, he had included a cosmological constant to keep the balance of the universe. When the new theory of the expansion of the universe based on the Doppler Effect came in the 1930s, he removed that constant and called it the biggest blunder of his life. That encouraged other scientists to believe that the unknown force of the expansion of the universe was more than the contracting force of gravity. This line of thinking gave a boost to the Big Bang theory, that the unknown expansion force must have come from the initial explosion during the Big Bang, and the gravity is only trying to slow it down.

But, when we explore the Lines of Space concept, we find that gravity is not the force causing compression of matter. On the other hand, it is the force generated in the space around a massive object due to the contraction of the medium of space into matter. Thus, there is no such requirement of an unknown force to counterbalance the gravity. So, it is evident that the presence of an unknown expansion force is also not correct as per the Lines of space concept. Thus, all the paths would have led to a concept of creation of the universe other than the Big Bang if only the existence of Ether had not been discarded in the nineteenth century. In the twenty first century, we must revisit the experiments done for proving the existence of a medium of space with more advanced technology available to us. That will give us more understanding of the universe.

I came out of my cogitations when the car swerved suddenly to the left to enter the gate of the Institute.

Chapter 20
The Leave Taking

Later that day, as we came out of the classroom, after saying goodbye to our instructors and other colleagues who had attended the course with us, Mr. Singh asked me, "Sir, what time will you be leaving tonight?" Being the last day, the course had wrapped up by noon as many of our classmates had come from far off places and wanted to return back to their hotels so they wouldn't have to rush to get ready for their evening flights back home.

"I'll be going tomorrow morning," I informed him.

"Why?"

"There are no direct flights to Chandigarh in the evening," I said.

"Okay. We shall be leaving in the evening," he said.

"Sir, it has been a good time here; we did the engine course, as well as learned about the universe," said Gaurav who was following us and heard our conversation.

"Yes, it was good to learn about the Lines of Space. Thanks," Mr. Singh also expressed his feelings.

"You are welcome," I said and noted that the van driver was waiting for us. As soon as the driver saw us, he cheerfully greeted us and started narrating the first-hand experience of the breakdown of his van that morning, and how he had to get it towed to a workshop. The conversation went on until we reached the guest house.

Both my colleagues started getting ready to go home. After about two hours, they departed for the airport with a promise to keep in touch, whereas I stayed back. I sat alone in my room, reviewing the events that had unfolded in the last five days, how an innocuous comment from Gaurav on the vastness of space started the whole story, and how I was able to convince them about the existence of the medium of space. Although I had to avoid many difficult topics of physics for the sake of simplicity, keeping in mind their knowledge of physics, but in the end it was satisfying to know that they preferred the reasonable approach of the Lines of Space than the implausible Big Bang for the creation of the universe.

Acknowledgment

I could not have written this book without the tremendous support from all the readers and the reviewers of the first book. Therefore, first of all, I want to thank them for reading *Lines of Space*, and giving very encouraging feedback. I want to thank my colleagues – Captain Sharat Kumar and Captain Sanjay Sharma who sat with me for hours discussing the concept of Lines of Space and talked me out of my negativities, motivated me, and believed in this concept. My shipmates – Captain Sunil Shetye, Sekhar M. Patkar, Anupam Sharma and Lovedeep Singh Aulakh propped me up. My classmates, whom I had not seen for the last twenty five years, gave encouragement by sending online messages; especially Ganga Shankar Dixit, Bharat Bhooshan Sharma, Manu Singh and Sanjay Kumar Jain; who were very encouraging.

I wish to thank two more of my colleagues; Anoop Abraham and Kanwar Singh, who listened uncomplainingly to my discourse during a training session that they attended with me in Mumbai, and in the process, inadvertently gave me the background and main characters for this book.

I want to thank my son Nikhil, daughter Priya and nephew Sandeep for spreading the word online about *Lines of Space*, and getting very good response from their friends.

I want to thank Author's Cave for making this book as error-free as possible and awarding their 'Seal of Excellence'. My thanks go specially to their editor Chameleon for working with me to make this book better. Thanks to Aidana Willowraven for design of the beautiful cover of this book.

And I want to thank all the readers for selecting and reading this book. I hope, you enjoyed reading it and found something in it to churn your thoughts about the creation of our Universe.

Author's Request for Feedback

Thank you for reading this book. Before you go, please rate this book and leave feedback as to whether you liked this book or not and why; on amazon.com or goodreads.com or any other platform, whichever you use. Although a lot of effort has gone into making this book as error-free as possible, if you find something wrong in this book, please convey to me, I shall be very grateful for your feedback and will make the amendments. If you have any questions related to the content of this book, you can post on **askdevinderdhiman@goodreads.com** or contact me on **http://linesofspace.webs.com**. I shall try my best to reply as soon as possible.

About the Author

The Author was born in 1965 in a small town near Chandigarh, in a middle class family. He completed his engineering degree from Marine Engineering and Research Institute in Kolkata. After graduation, he worked as an engineer on a ship. In 1999, he became Chief Engineer. He works on ships for about six months every year and devotes the remaining time to his interest of physics. He published his first book *'Lines of Space'* in 2012. He lives with his family in Panchkula – a beautiful town in north India.

*****

Other Book By The Author

Lines of Space- *Source of Fundamental Forces and Constituent of all Matter in the Universe*

Available in print on Amazon.com and flipkart.com

Available as an ebook at http://amzn.to/15Yg5JY, smashwords.com and at flipkart.com

Declaration

All the discussions with the characters of this story are the Author's own deliberations. The trip to Mumbai provided only the background for the story. There was only a passing discussion regarding the Big Bang with colleagues over there, not in details as presented in this book.

www.ingramcontent.com/pod-product-compliance
Lightning Source LLC
Chambersburg PA
CBHW051717170526
45167CB00002B/694